# Springer Series in Materials Science

## Volume 179

*Series Editors*



For further volumes:
http://www.springer.com/series/856

The Springer Series in Materials Science covers the complete spectrum of materials physics, including fundamental principles, physical properties, materials theory and design. Recognizing the increasing importance of materials science in future device technologies, the book titles in this series reflect the state-of-the-art in understanding and controlling the structure and properties of all important classes of materials.

Yongkang Zhang · Jinzhong Lu
Kaiyu Luo

# Laser Shock Processing of FCC Metals

## Mechanical Properties and Micro-Structural Strengthening Mechanism

Springer

Yongkang Zhang
School of Mechanical Engineering
Southeast University
Nanjing City
People's Republic of China

Jinzhong Lu
Kaiyu Luo
School of Mechanical Engineering
Laser Technology Institute
Jiangsu University
Zhenjiang City
People's Republic of China

ISSN 0933-033X
ISBN 978-3-642-35673-5 ISBN 978-3-642-35674-2 (eBook)
DOI 10.1007/978-3-642-35674-2
Springer Heidelberg New York Dordrecht London

Library of Congress Control Number: 2013931648



Printed on acid-free paper

Springer is part of Springer Science+Business Media (www.springer.com)

# Preface

Laser shock processing, or laser shock peening (LSP), is the process of hardening or strengthening metal using a powerful laser. It can generate a layer of residual compressive stress on a surface of metallic materials and alloys that is four times deeper than that attainable from conventional shot peening treatments (shot peening), which has been successfully applied to improve fatigue performance of metallic components.

In the past three decades, LSP has been widely and intensively investigated over 200 scientific papers and reports. Most studies and investigations have been based on experimental approaches, influences of LSP on mechanical properties and in particular fatigue lives of metallic materials and alloys. Many researches have been focusing on analytical models and dynamic finite element models (FEM), to simulate the distribution of three-dimensional residual stresses in relation to materials properties, component geometry, laser sources, and LSP parameters in the last decade. LSP is also an effective surface treatment and post-processing method to eliminate tensile residual stress in the surface layer of metallic material and its weldment in order to improve their mechanical properties and tensile performances.

In this book, we take the face-centered cubic metals (FCC metals, including aluminum alloy and austenitic stainless steel) and stainless steel weldment as the studied objects. The aim of this book is to provide some foundational researches on the macro-property, micro-structure evolution, and plastic deformation induced by massive LSP impacts. These researches can provide some scientific insights into the industry application of LSP technology. Some different topics are involved, i.e., surface integrity and fatigue lives of FCC metals after LSP with different processing parameters, tensile property, and fractural morphology of FCC metals by LSP under different strain-rates, grain refinement mechanism based on the micro-structure evolution, and corrosion behaviors after multiple LSP impacts. Special attentions have been paid to the effects of LSP on mechanical properties and tensile performance of stainless steel weldment.

For a better understanding on the effects of LSP on the macro-properties and micro-structures, and in order to obtain the appropriate LSP processing criterion by addressing the various factors mentioned above, a lot of LSP experiments are carried out, and finite element simulation based on mechanistic modeling is currently recognized as an effective tool to analyze the distribution of residual stress. Some influential parameters associated with LSP are evaluated for the purpose of characterizing LSP processes. In particular, different methods of using LSP, such as one-sided, two-sided, and multiple LSP impacts on the FCC metals are elaborated in detail.

The research work from which this book arises was carried out at Laser Technology Institute in Jiangsu University supported by the National Natural Science Foundation of China (Grant Nos: 50735001, 51275220, 50675089, 50705038, 51105179). The work was based on research projects on evaluation and characterization of LSP for aerospace and engineering applications. Professor Yongkang Zhang at Southeast University finished Chaps. 1 and 5, Dr. Jinzhong Lu at Jiangsu University finished Chaps. 1–4 and 9, and Dr. Kaiyu Luo at Jiangsu University finished Chaps. 6–8. The authors would like to thank their colleagues and friends for useful discussions and help in the preparation of this book. The authors are particularly grateful to Dr. Fengze Dai, Dr. Lei Zhang, and Prof. Xinmin Luo due to their contribution to this book.

Finally, Dr. Jinzhong Lu would like to thank his family, for their love, understanding, and assistance over the years.

Nanjing City, People's Republic of China
Zhenjiang City, People's Republic of China

Yongkang Zhang
Jinzhong Lu
Kaiyu Luo

# Contents

# Chapter 1
# General Introduction

**Abstract** This chapter presents the laser shock processing (LSP) process, recent development of LSP on alloys and metallic materials, typical applications of LSP and the scope of this book.

## 1.1 Laser Surface Treatment and Laser Shock Processing

Laser surface treatment is a flexible way of effectively protecting component and tool surfaces from wear and corrosion. Various surface treatment techniques can prolong tool life and improve component, tool or die performance.

Laser surface treatment is a novel and potential subject of considerable interest at present because it seems to offer the chance to save strategic materials or to allow improved components with idealized surfaces and bulk properties. Laser surface treatment shows many advantages compared to conventional techniques, e.g. a high flexibility with respect to the processed geometries or the possibility for a simple integration into existing production lines. Especially when only a small part of the surface of a workpiece shall be treated the laser process should be preferred.

Laser surface treatment can cause the changes in surface properties of alloys and metallic materials by generating temperature gradients, phase changes or mechanical influences. The most important processes are shown as following:

- *Laser transformation hardening* It is used for steel and cast iron. Heating by the laser above the austeniting temperature causes an $\alpha$-$\gamma$ transition of the material in the heated surface layer. Because of the high power density of the laser beam high temperature gradients are caused in the material which induces fast quenching when the laser beam moves ahead. This fast quenching causes the forming of hard surface layers consisting of martensite.
- *Laser remelting* In remelting the laser parameters are selected in a way that melting of the upper surface layer is occurring. There are several effects of the

Y. Zhang et al., *Laser Shock Processing of FCC Metals*,
Springer Series in Materials Science 179, DOI: 10.1007/978-3-642-35674-2_1,

melting which can be utilized. For example inclusions can be vaporized or dissolved, the grain size can be adjusted and the hardness can be increased. Because of the flow of the produced melt the surface quality gets worse and usually additional processing is required.

- *Laser annealing* Laser annealing process is very similar to transformation hardening. In this process a material with a high martensite fraction is heated in order to dissolve a part of the martensite and thus to reduce the hardness.
- *Laser shock processing* Shock hardening occurs when laser pulses with a duration in the ns range are applied. In this case shock waves are induced which cause a kind of mechanical deformation connected to an increase of the hardness. The involved mechanisms are therefore similar to those occurring during cold working.

Figure 1.1 shows laser surface treatment processes in relation to other laser processes used in production engineering. It can be seen that laser surface treatment requires a relatively high energy laser beam. The interaction time is determined by the required heating and quenching rates, which is higher than that used for welding or cutting.

Among these laser surface treatment techniques, laser shock processing (LSP), also known as laser shock peening or laser peening, is a novel surface modification technique, which is successfully applied to improve the fatigue performance and mechanical properties of alloys and metallic materials due to the fact that LSP can impart a layer of compressive residual stress in the surface that is four times deeper than that attainable from conventional shot peening. After the treatment, the fatigue strength and fatigue life of metallic materials can be increased significantly

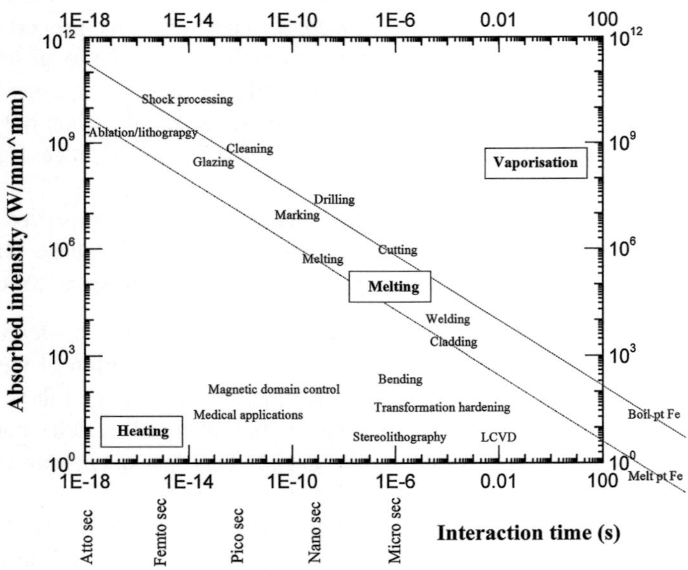


**Fig. 1.1** Overview over laser surface treatment processes

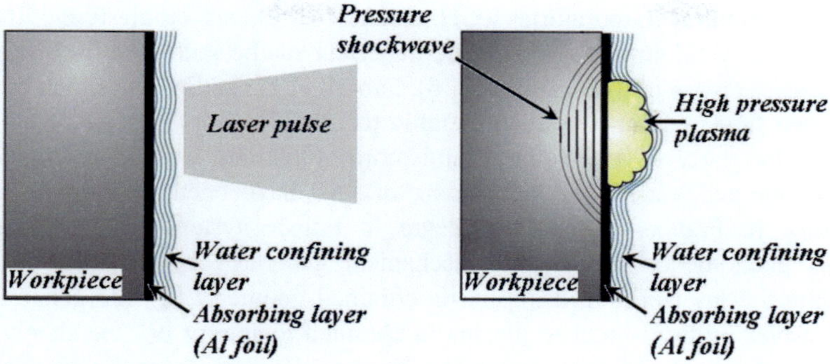


**Fig. 1.2** Schematic principle of laser shock processing

due to the presence of compressive residual stresses in the alloys and metallic materials.

The LSP process utilizes high energy laser pulses (several GW/cm$^2$) fired at the surface of a metal covered by two layers, namely an absorbing layer and a water confining layer. When a laser pulse with sufficient intensity passes through the transparent confining layer and hits the surface of the material, the absorbent material vaporizes and forms a plasma. The plasma continues to strongly absorb the laser energy until the end of the energy deposition. The rapidly expanding plasma is trapped between the sample and the transparent confining layer, creating a high surface pressure, which propagates into the material as a shockwave. When the pressure of the shockwave exceeds the dynamic yield strength of the metal, it produces plastic deformation in the near-surface of the metal. The LSP principle is schematically shown in Fig. 1.2.

The LSP technology has the following advantages: (1) less surface roughening compared to the conventional shot peening; (2) no embedded particles; (3) strengthening right into corners where shot could not reach; (4) no material to recycle, collect, grade and clean, as there is with shot peening; and (5) flexible laser pulse beam which can be adjusted and controlled in real time.

## 1.2 Recent Development of Laser Shock Processing on Alloys and Metallic Materials

A pulsed laser beam with a pulse width in the level of nanosecond can induce the shock wave, which was discovered and developed in the early 1960s for the first time [1, 2]. Subsequently, a lot of research focused on the improvement in mechanical properties (including micro-hardness, residual stress, fatigue life and yield strength) of alloys and metallic materials induced by LSP around the world. Anderholm et al. established the confined mode for an improved LSP process in 1968 [3]. Clauer et al. carried out LSP for the fastener holes in 1968–1981 at

Battelle–Columbus Laboratories (OH, USA) [4]. Banas et al. found that the hardness and yield strength of metallic materials can be increased by LSP due to high density arrays of dislocations [5, 6]. Fabbro et al. studied the shock breakout at the rear face of laser-irradiated metallic targets [7].

Since the early of 1990s, more and more systematic studies on engineering applications and underlying mechanism of LSP have been performed in other countries. In France, Fabbro and Peyre in France systematically studied the physics phenomenon and action mechanism generated by laser shock wave, including the laser-produced plasma in confined geometry [8], the generation of shock waves by laser-induced plasma in confined geometry [9], the shock waves from a water-confined laser-generated plasma [10], the laser-driven spallation process by the velocity interferometer system for any reflector interferometry technique [11, 12], laser-driven shock waves in stainless steels [13], and the wavelength dependent of laser shock-wave generation in the water-confinement regime [14]. In China, Zhang and Yu investigated the effects of LSP on 2024-T62 aluminum alloy [15], and results showed that LSP can increase the fatigue life and decrease the fatigue crack growth rates of 2024-T62 aluminum alloy, which results from the combinations of the surface compressive residual stress, reduced surface roughness, and increased dislocation density induced by the laser shock waves. In addition, they also proposed a novel measurement method to predict the effect of LSP on 2024-T6 aluminum alloy by comparing the surface toughness of the shocked region with that of the unshocked surface [16]. In Japan, Sano et al. developed the LSP of water-immersed material to improve the surface residual stress of metal components. The process changes the stress field from tensile to compressive by means of impulsive pressure of laser-induced plasma generated through the ablative interaction of the intense laser pulse with the material [17].

With the rapid development of the laser equipment, LSP is attracting comprehensive attentions of more and more researchers in the field of surface modification due to high-pressure (in the scale of GPa), high-energy (peak power is more than 1 GW), ultrafast (several tens nanoseconds) and ultra-high strain rate (more than $10^7$ $s^{-1}$) of the induced laser shock wave [5, 6]. There are many researchers who have been focusing on surface integrity, mechanical properties, microstructures, fatigue life of alloys and metallic materials induced by LSP. Cheng and Ye et al. at Purdue University systematically investigated fatigue lives, mechanical properties and microstructures of A6160 Aluminum alloy [18], SUS304 SS [19], copper [20], NiTi shape memory alloy [21] and AISI 4140 steel [22] in the elevated temperature, room temperature and low temperature before and after LSP, respectively. They found that LSP can refine and nanocrystallization the coarse grain in the surface layers of alloys and metallic materials in the above three temperatures, which can result in the improvement of mechanical properties and fatigue performance of alloys and metallic materials in the elevated temperature. Zhang et al. at Jiangsu University found that LSP changed the elastic properties of 2024-T62 aluminum alloy due to the distribution of microstructural changes induced by LSP [23]. Research results of Lu et al. at Jiangsu University showed that the values of the nano-hardness, elastic modulus and surface residual stress in the laser-shocked region and the laser-affected region were obviously improved compared to those in the non-shocked region [24]. Gomez-Rosas et al. studied the effects of LSP on wear and friction of 6061-T6 aluminum alloy, results showed that wear rate decreases as pulse density increases [25]. Lu et al. at Jiangsu University addressed the effects of multiple LSP impacts with different pulse energy on mechanical properties and wear behaviors of AISI 8620 steel [26]. Multiple LSP impacts can remarkably improve the wear resistance of AISI 8620 steel. The wear process of the unpolished sample subjected to multiple LSP impacts can be described as follows: the wear rate was big at the beginning of sliding dry wear, but then decreased after the micro-indention in the sample surface was polished to the disappear. Multiple LSP impacts on AISI 8620 steel had dual-function: the refinement of coarse grains in the near-surface region by dislocation movement and dispersion strengthening of C atoms which cut cementite and diffused into the ferrite by moving dislocations [27].

The change of residual stress on the surface and in the depth direction attracted more attention in the field of LSP. The effects of LSP parameters [28–34], including absorbing layer, pulse width, laser spot diameter, overlapping rate, pulse energy, wavelength and the time of LSP impact, on residual stress on the surface and in the depth direction were widely studied. Many alloys and metallic materials presented clear improvements in fatigue life after LSP treatment. The beneficial effects of LSP may originate from compressive residual stresses on the surface and in depth direction, which delayed the propagation of surface fatigue cracking and the growth of fatigue cracking in depth direction. Investigations on several different aspects of the fatigue behaviors, such as fatigue life, fatigue strength and fretting fatigue, were reported [35–42].

LSP can significantly improve the corrosion resistance and wear performances of alloys and metallic materials. Amar et al. at CNRS-Université de Bourgogne researched the corrosion behavior of AA2050-T8 after polishing and then LSP treatment using the electrochemical microcell technique and the SVET. Results revealed that LSP increased the pitting potential [43]. Trdan et al. in Faculty of Mechanical Engineering at University of Ljubljana studied corrosion resistance of AA6082-T651 aluminum alloy after laser shock peening by means of cyclic polarization and ElS methods, and they found that LSP can reduce pitting and completely prohibit the initiation of large surface pits to enhance corrosion resistance of AA6082 alloy [44, 45]. Zhang and Lu et al. investigated the effects of LSP on the stress corrosion cracking (SCC) susceptibility of AZ31B magnesium (Mg) alloy, and the SCC test in 1 wt % NaOH solution showed that LSP retarded the SCC initiation and growth in AZ31B Mg alloy [46].

There are some new developments in the plasma-related phenomenon and mechanism induced by laser shock wave. Wu and Shin et al. in the Center for Laser-Based Manufacturing at Purdue University paid attention to the plasma induced by laser shock wave in the past decade [47]. They proposed a self-closed numerical model which can simulate the laser pulse transmission through the breakdown plasma generated in water during LSP. This model can predict reasonably good experimental results. In addition, they also developed a physics-based predictive model to measure the early-stage plasma pressure and front propagation and the late-stage (t > 30 ns) plasma temperature and electron number density during LSP, which is very useful for the fundamental laser plasma study and relevant laser applications [48]. Wu et al. at Chinese Academy of Physics proposed a coupling pressure analytical model, in which the material constitutive models of confined layers and target material are considered, and this model can predict the plasma pressure profile at the surface of metallic target [49]. Thorslund et al. at University of Central Florida developed some mathematical models to calculate the temperatures, pressures and stresses during LSP for time-modulated (ramp-up, ramp-down and rectangular) laser pulses [50].

## 1.3 Typical Applications of Laser Shock Processing

LSP attracts strong interest in the field of commercialization application. Since two important patents were first issued in 1974 and 1983, LSP has been gradually used in industry. From 1996, the General Electric Company alone applied for a large number US patents based on laser shock processing.

In the aerospace industry, laser shock processing is an effective methods to improve the mechanical properties and fatigue lives of aerospace key products, such as turbine blades (as shown in Fig. 1.3) [51] and rotor components [52, 53], discs, gear shafts [54] and bearing components [55]. Laser shock processing could also be applied to strengthen fastener holes in cover parts. General Electric Aircraft Engines in the USA treated the leading edges of turbine fan blades [53] in F101-GE-102 turbine for the Rockwell B-1B bomber by laser shock processing in 1997, which enhanced fan blade durability and resistance to foreign object damage (FOD) without harming the surface finish [53]. Protection of turbine engine components against FOD [51] is a key priority of the US Air Force. In addition, it was reported that laser peening would be applied to treat engines used in the Lockheed Martin F-16C/D [56].

LSP Technologies Incorporation (Dublin, OH) also recently commissioned the ManTech Laser Shock Peening Manufacturing Cell (LSPMC). From 2004, LSP Technologies laser peened the airfoils on the Pratt and Whitney F119, 4th stage IBR that was flown on the F/A-22 Raptor aircraft. Implementation of laser shock processing increased the notched fatigue strength of IBR airfoils above the 55 ksi fatigue strength design criteria. Figure 1.3 shows an improvement of the notched fatigue strength of IBR airfoils [57]. The application of laser shock processing to the F119 IBR has reduced maintenance costs and eliminated the need for a costly redesign, estimated to be greater than $10 M (Fig. 1.4).

Laser shock processing can obviously improve the fatigue strength of the damaged blades which is equal to or better than that of undamaged blades [58], as shown in Fig. 1.5.

Laser shock processing has been shown to provide more than 2.5 times lifetime enhancement against fatigue failure for coupons replicating T-45 arrestment hook

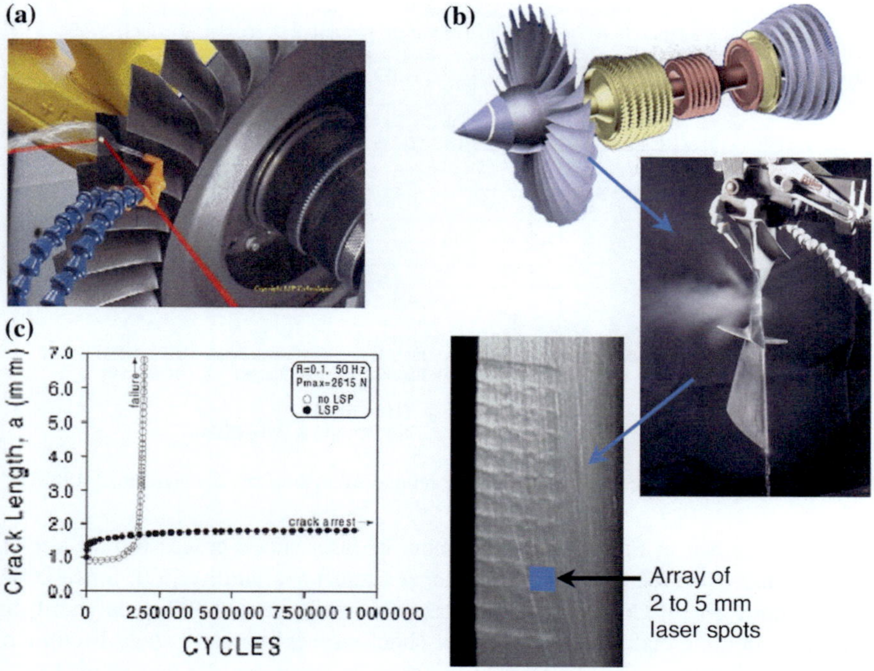


**Fig. 1.3** **a** Laser shock processing of F119 vane-integrated disk and blade. **b** Blade. **c** Crack length vs. cycles (a vs. N) data for baseline and LSP samples

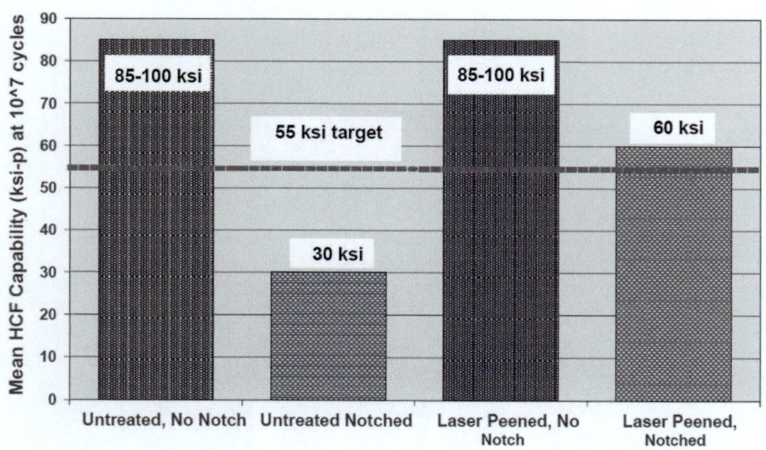


**Fig. 1.4** The effect of laser shock processing on F119 IBR fatigue life. A 0.050-inches deep EDM notch was used on the notched airfoils [57]

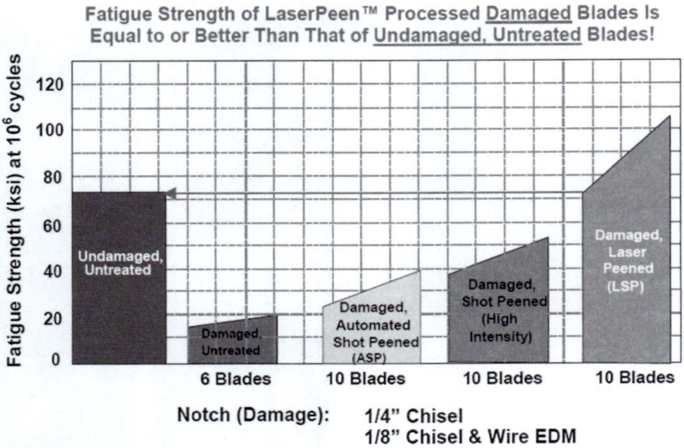


**Fig. 1.5** Laser shock processing improved the fatigue strength of the damaged blades [58]

shanks (as shown in Fig. 1.6). Employment of laser shock processing on aircraft would reduce maintenance costs and add to aircraft availability [59]. Since 2002, Metal Improvement Company (MIC) has processed over 35,000 wide-chord fan blades for commercial aircraft, and laser shock processing can extend lifetime of new and used Being 777 blades by more than 20 times (as shown in Fig. 1.6).

Laser shock processing without protective coating (LSPwC) was also developed and the practical effects (SCC and fatigue prevention). This technology has been applied to Japanese nuclear power reactors (BWRs and PWRs) as preventive maintenance against SCC since 1999 [60, 61] (Fig. 1.7).

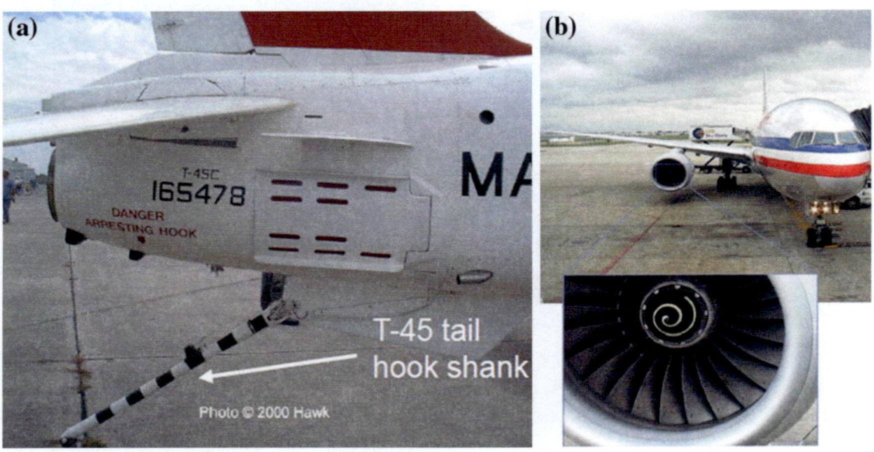


**Fig. 1.6** **a** The replicating T-45 arrestment hook shanks treated by laser shock processing. **b** The treated being 777 blades by laser shock processing

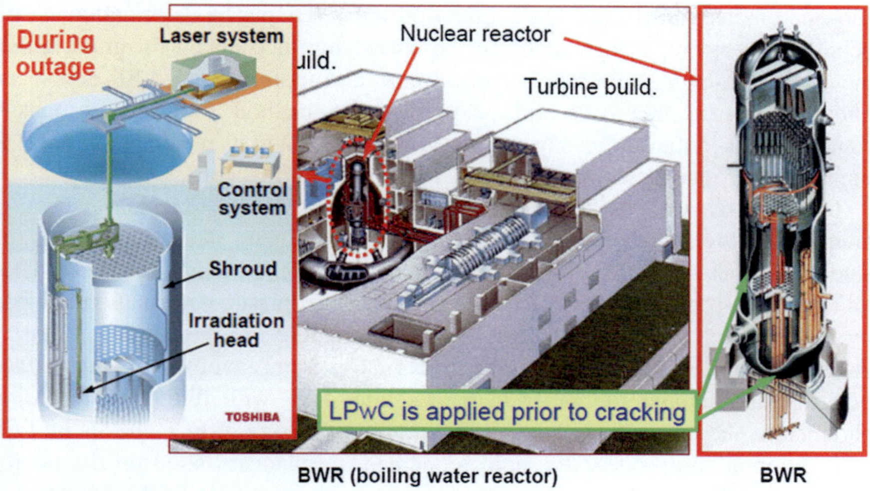


**Fig. 1.7** Laser shock processing without protective coating was applied to retard stress corrosion cracking [60, 61]

## 1.4 Scope of the Book

The enhanced properties and microstructural changes in alloys and metallic materials induced by LSP have widely been presented by many researchers. Since the early 2000s, a number of researchers have conducted many experiments to elaborate the effects of the confined interaction mode and the factors influencing the laser pulse during an LSP process.

The main effect of LSP is to introduce high-level compressive residual stress and refine the coarse grain in the surface layer that can result in an improvement in the mechanical performance of alloys and metallic materials such as fatigue performance and corrosion resistance. The distribution of residual stress in the LSPed (treated by laser shock processing) metallic component strongly depends on the generation and propagation of laser shock wave and its interaction with the metallic component, for example the geometry and material properties in relation to a single or multiple LSP impacts.

Typical LSP parameters with a confined mode include the laser power density, shock time, overlapping rate and laser spot size. The inappropriate processing parameter set can induce unexpected tensile residual stresses, which will be very detrimental to the mechanical performance of the metallic component. Although many researches have paid enough attentions to the surface integrity (including surface roughness, residual stress, micro-hardness), mechanical properties, fatigue performance and corrosion resistance, little literature has focused on the correlation between macro-mechanical properties and micro-structures of alloys and metallic materials. Nevertheless, up to now, a clear scenery between micro-structure and macro-property of the refined surface layer, especially formation of sub-micrometer grains from coarse grains during severe plastic deformation, is still pending. Therefore, a basic study of the underlying mechanism for grain refinement by ultra-high strain-rate becomes more and more crucial. LSP is also an effective surface treatment and post-processing method to eliminate tensile residual stress in the surface layer of metallic material and its weldment in order to improve their mechanical properties and tensile performances.

In this book, we take the face-centered cubic metals (FCC metals, including aluminum alloy and austenitic stainless steel) and stainless steel weldment as the studied objects. The aim of this work is to provide some foundational researches on the macro-property, micro-structure evolution and plastic deformation induced by LSP. These researches can provide some scientific insights into the industry application of LSP technology. Some different topics are involved, i.e. the surface integrity and fatigue life of FCC metals after LSP with different processing parameters, the tensile property and fractural morphology of FCC metals by LSP with different strain-rates, the grain refinement mechanism based on the micro-structure evolution, and the corrosion behaviors after multiple LSP impacts.

For a better understanding on the effects of LSP on the macro-properties and micro-structures, and in order to obtain the appropriate LSP processing criterion by addressing the various factors mentioned above, a lot of LSP experiments are carried out, and finite element simulation based on mechanistic modeling is currently recognized as an effective tool to analyze the distribution of residual stress in the approach, if the simulation procedures have been well calibrated and validated by the experimental data. Some influential parameters associated with LSP are evaluated for the purpose of characterizing LSP processes. In particular, different methods of using LSP, such as one-sided, two-sided and multiple LSP impacts on the FCC metals are elaborated in detail.

The outline of this book is described as follows.

- Chapter 2 presents a comprehensive literature review of the physical and mechanical mechanisms of LSP on metallic materials, which have been investigated in the past 30 years. In particular, more attentions have been focused on the effects of LSP on mechanical properties of LY2 Al alloys are also highlighted.
- Chapter 3 presents the effects of strain rate on the tensile properties and fracture morphologies of LY2 Al alloy subjected to LSP. Special attentions are paid to the effects of LSP processing parameters on the fatigue life and fracture morphology. In addition, the influence mechanisms of different LSP parameters on Al alloy are systematically revealed.
- Chapter 4 presents the effects of the multiple LSP impacts on the residual stresses in the depth direction, focusing especially on the change of micro-structure in different layers in the treated sample by multiple LSP impacts, and clearly reveals the underlying refinement mechanism of plastic deformation after multiple LSP impacts on LY2 Al alloy.
- Chapter 5 introduces a novel visual inspection and control method to evaluate the surface qualities of Al alloy by means of three parameters, i.e., surface

roughness, surface residual stress, and hollow depth, and also presents the effects of the surface qualities of LSP zones upon the fatigue life of samples and the optimum conditions of LSP.

- Chapter 6 presents the effects of a single LSP impact on the nano-hardness, elastic modulus, residual stress, and phase transformation of AISI 304 SS. Further studies of different LSP paths on the mechanical properties and fracture morphology of the laser welded AISI 304 SS joint are presented and discussed.
- Chapter 7 investigates the effects of massive LSP impacts on the SCC behavior of AISI 304 SS and highlights the distribution of surface residual stress, the SCC initiation and micro-structure on the top surface of three types of U-bend samples. The emphasis is placed on the mechanism of massive LSP impacts on the corrosion resistance of AISI 304 SS.
- Chapter 8 describes the micro-structural evolution and grain refinement process of AISI 304 SS subjected to multiple LSP impacts by means of cross-sectional optical microscopy and transmission electron microscopy observations, and reveals the plastic strain-induced grain refinement mechanism of FCC materials with very low stacking fault energy.
- Chapter 9 figures out effects of LSP on electrochemical corrosion resistance of AISI 304 SS weldments after cavitation erosion. Through the electrochemical test, AISI 304 SS weldments without and with LSP impacts after cavitation erosion are studied. Influence mechanism of LSP on electrochemical corrosion resistance of AISI 304 SS weldments after cavitation erosion is also revealed.

## References

1. Askar, C. A., & Moroz, E. M. (1963). Pressure on evaporation of matter in a radiation beam. *Journal of Experimental and Theoretical Physics Letters, 16*, 1638–1644.
2. White, R. M. (1963). Elastic wave generation by electron bombardment or electromagnetic wave absorption. *Journal of Applied Physics, 34*, 2123–2124.
3. Anderholm, N. C. (1970). Laser-generated stress wave. *Applied Physics Letters, 16*(3), 113–115.
4. Clauer, A.H., Holbrook, J.H., Fairand, B.P. (1981). Effects of laser induced shock waves on metals. In: M.A. Meyers, L.E. Murr (Ed.), *shock waves and high-strain-rate phenomena in metals* (pp. 675–703). New York: Plenum Publishing Corporation.
5. Banas, G., Elsayed-ali, H. E., Lawrence, F. V., & Rigsbee, J. M. (1990). Laser shock induced mechanical and microstructural modification of welded maraging steel. *Applied Physics, 67*, 2380–2384.
6. Banas, G., Lawrence, F. V., Rigsbee, J. M., & Elsayed-ali, H. E. (1990). Laser shock hardening of welded maraging steel. *Surface Engineering, 67*, 280–290.
7. Cottet, F., Marty, L., Hallouin, M., & Romain, J. P. (1988). Two-dimensional study of shock breakout at the rear face of laser-irradiated metallic targets. *Journal of Applied Physics, 64*(9), 4473–4476.
8. Fabbro, R., Fournier, J., Ballard, P., Devaux, D., & Virmont, J. (1990). Physical study of laser-produced plasma in confined geometry. *Journal of Applied Physics, 68*(2), 775–784.

9. Devaux, D., Fabbro, R., Tollier, L., & Bartnicki, E. (1993). Generation of shock waves by laser-induced plasma in confined geometry. *Journal of Applied Physics, 74*(4), 2268–2273.
10. Berthe, L., Fabbro, R., Peyre, P., Tollier, L., & Bartnicki, E. (1997). Shock waves from a water-confined laser-generated plasma. *Journal of Applied Physics, 82*(6), 2826–2832.
11. Tollier, L., Fabbro, R., & Bartnicki, E. (1998). Study of the laser-driven spallation process by the velocity interferometer system for any reflector interferometry technique. I. Laser-shock characterization. *Journal of Applied Physics, 83*(3), 1224–1230.
12. Tollier, L., & Fabbro, R. (1998). Study of the laser-driven spallation process by the VISAR interferometry technique II: experiment and simulation of the spallation process. *Journal of Applied Physics, 83*(3), 1231–1237.
13. Peyre, P., Berthe, L., Scherpereel, X., Fabbro, R., & Bartnicki, E. (1998). Experimental study of laser-driven shock waves in stainless steels. *Journal of Applied Physics, 84*(11), 5985–5992.
14. Berthe, L., Fabbro, R., Peyre, P., & Bartnicki, E. (1999). Wavelength dependent of laser shock-wave generation in the water-confinement regime. *Journal of Applied Physics, 85*(11), 7552–7555.
15. Zhang, H., & Yu, C. Y. (1998). Laser shock processing of 2024–T62 aluminum alloy. *Materials Science and Engineering A, 257*, 322–327.
16. Zhang, Y. K., Zhang, S. Y., Zhang, X. R., Cai, L., Yang, J. C., & Ren, N. F. (1997). Investigation of the surface qualities of laser shock-processed zones and the effect on fatigue life of aluminum alloy. *Surface & Coatings Technology, 92*(1–2), 104–109.
17. Sano, Y., Mukai, N., Okazaki, K., & Obata, M. (1997). Residual stress improvement in metal surface by underwater laser irradiation. *Nuclear Instruments & Methods in Physics Research, Section B: Beam Interactions with Materials and Atoms, 121*, 432–436.
18. Ye, C., Liao, Y. C., & Cheng, G. J. (2010). Warm laser shock peening driven nanostructures and their effects on fatigue performance in aluminum alloy 6160. *Advanced Engineering Materials, 12*(4), 291–297.
19. Ye, C., Suslov, S., Lin, D., & Cheng, G. J. (2012). Deformation-induced martensite and nanotwins by cryogenic laser shock peening of AISI 304 stainless steel and the effects on mechanical properties. *Philosophical Magazine, 92*(11), 1369–1389.
20. Ye, C., & Cheng, G. J. (2010). Effects of temperature on laser shock induced plastic deformation: the case of copper. *Journal of Manufacturing Science and Engineering, 132*, 061009.
21. Ye, C., Suslov, S., Fei, X. L., & Cheng, G. J. (2011). Bimodal nanocrystallization of NiTi shape memory alloy by laser shock peening and post-deformation annealing. *Acta Materialia, 59*, 7219–7227.
22. Ye, C., Suslov, S., Kim, B. J., Stach, E. A., & Cheng, G. J. (2011). Fatigue performance improvement in AISI 4140 steel by dynamic strain aging and dynamic precipitation during warm laser shock peening. *Acta Materialia, 59*, 1014–1025.
23. Zhang, Y. K., Zhang, X. R., Wang, X. D., Zhang, S. Y., Gao, C. Y., Zhou, J. Z., et al. (2001). Elastic properties modification in aluminum alloy induced by laser-shock processing. *Materials Science and Engineering A, 297*(1–2), 138–143.
24. Lu, J. Z., Zhang, L., Feng, A. X., Jiang, Y. F., & Cheng, G. G. (2009). Effects of laser shock processing on mechanical properties of Fe–Ni alloy. *Materials and Design, 30*(9), 3673–3678.
25. Sánchez-Santana, U., Rubio-González, C., Gomez-Rosas, G., Ocaña, J. L., Molpeceres, C., Porro, J., et al. (2006). Wear and friction of 6061–T6 aluminum alloy treated by laser shock processing. *Wear, 260*(7–8), 847–854.
26. Lu, J. Z., Luo, K. Y., Dai, F. Z., Zhong, J. W., Xu, L. Z., Yang, C. J., et al. (2012). Effects of multiple laser shock processing (LSP) impacts on mechanical properties and wear behaviors of AISI 8620 steel. *Materials Science and Engineering A, 536*, 57–63.
27. Lu, J. Z., Zhong, J. W., Luo, K. Y., Zhang, L., Dai, F. Z., Chen, K. M., et al. (2011). Microstructural strengthening mechanism of multiple laser shock processing impacts on AISI 8620 steel. *Materials Science and Engineering A, 528*(19–20), 6128–6133.

28. Cellard, C., Retraint, D., François, M., Rouhaud, E., & Saunier, D. L. (2012). Laser shock peening of Ti-17 titanium alloy: Influence of process parameters. *Materials Science and Engineering A, 532*, 362–372.
29. Rubio-González, C., Gomez-Rosas, G., Ocaña, J. L., Molpeceres, C., Banderas, A., Porro, J., et al. (2006). Effect of an absorbent overlay on the residual stress field induced by laser shock processing on aluminum samples. *Applied Surface Science, 252*(18), 6201–6205.
30. Hu, Y. X., & Yao, Z. Q. (2008). Overlapping rate effect on laser shock processing of 1045 steel by small spots with Nd: YAG pulsed laser. *Surface & Coatings Technology, 202*(8), 1517–1525.
31. Zhang, L., Luo, K. Y., Lu, J. Z., Zhang, Y. K., Dai, F. Z., & Zhong, J. W. (2011). Effects of laser shock processing with different shocked paths on mechanical properties of laser welded ANSI 304 stainless steel joint. *Materials Science and Engineering A, 528*(13–14), 4652–4657.
32. Gomez-Rosas, G., Rubio-Gonzalez, C., Ocaña, J. L., Molpeceres, C., Porro, J. A., Chi-Moreno, W., et al. (2005). High level compressive residual stresses produced in aluminum alloys by laser shock processing. *Applied Surface Science, 252*(4), 883–887.
33. Gomez-Rosas, G., Rubio-Gonzalez, C., Ocaña, J. L., Molpeceres, C., Porro, J. A., Morales, M., et al. (2010). Laser shock processing of 6061–T6 Al alloy with 1064 nm and 532 nm wavelengths. *Applied Surface Science, 256*(20), 5828–5831.
34. Dorman, M., Toparli, M. B., Smyth, N., Cini, A., Fitzpatrick, M. E., & Irving, P. E. (2012). Effect of laser shock peening on residual stress and fatigue life of clad 2024 aluminium sheet containing scribe defects. *Materials Science and Engineering A, 548*, 142–151.
35. Gerland, M., & Hallouin, M. (1994). Effect of pressure on the microstructure of an austenitic stainless steel shock-loaded by very short laser pulses. *Journal of Materials Science, 29*, 345–351.
36. Montross, C. S., Florea, V., Brandt, M., & Swain, M. V. (2000). Subsurface properties of laser peened 6061–T6 Al weldments. *Surface Engineering, 16*, 116–121.
37. Montross, C. S., Brandt, M., & Swain, M. V. (2001). Self-limiting hardness changes in laser peened 6061–T6 aluminum. *Surface Engineering, 17*(6), 477–482.
38. Maawad, E., Sano, Y., Wagner, L., Brokmeier, H. G., & Genzel, C. (2012). Investigation of laser shock peening effects on residual stress state and fatigue performance of titanium alloys. *Materials Science and Engineering A, 536*(28), 82–91.
39. Yang, J. M., Her, Y. C., Han, N., & Clauer, A. (2001). Laser shock peening on fatigue behavior of 2024–T3 Al alloy with fastener holes and stopholes. *Materials Science and Engineering A, 298*(1–2), 296–299.
40. Chahardehi, A., Brennan, F. P., & Steuwer, A. (2010). The effect of residual stresses arising from laser shock peening on fatigue crack growth. *Engineering Fracture Mechanics, 77*(11), 2033–2039.
41. Rubio-González, C., Felix-Martinez, C., Gomez-Rosas, G., Ocaña, J. L., Morales, M., & Porro, J. A. (2011). Effect of laser shock processing on fatigue crack growth of duplex stainless steel. *Materials Science and Engineering A, 528*(3), 914–919.
42. Rubio-González, C., Ocaña, J. L., Gomez-Rosas, G., Molpeceres, C., Paredes, M., Banderas, A., et al. (2004). Effect of laser shock processing on fatigue crack growth and fracture toughness of 6061–T6 aluminum alloy. *Materials Science and Engineering A, 386*(1–2), 291–295.
43. Amar, H., Vignal, V., Krawiec, H., Josse, C., Peyre, P., Silva, S. N., et al. (2011). Influence of the microstructure and laser shock processing (LSP) on the corrosion behaviour of the AA2050-T8 aluminium alloy. *Corrosion Science, 53*(10), 3215–3221.
44. Trdan, U., & Grum, J. (2012). Evaluation of corrosion resistance of AA6082-T651 aluminium alloy after laser shock peening by means of cyclic polarisation and ElS methods. *Corrosion Science, 59*, 324–333.
45. Krawiec, H., Vignal, V., Amar, H., & Peyre, P. (2011). Local electrochemical impedance spectroscopy study of the influence of ageing in air and laser shock processing on the

micro-electrochemical behaviour of AA2050-T8 aluminium alloy. *Electrochimica Acta, 56*(26), 9581–9587.
46. Zhang, Y. K., You, J., Lu, J. Z., Cui, C. Y., Jiang, Y. F., & Ren, X. D. (2010). Effects of laser shock processing on stress corrosion cracking susceptibility of AZ31B magnesium alloy. *Surface & Coatings Technology, 204*(24), 3947–3953.
47. Wu, B. X., & Shin, Y. C. (2006). Laser pulse transmission through the water breakdown plasma in laser shock peening. *Applied Physics Letters, 88*, 041116.
48. Wu, B. X. (2008). High-intensity nanosecond-pulsed laser-induced plasma in air, water, and vacuum: a comparative study of the early-stage evolution using a physics based predictive model. *Applied Physics Letters, 93*, 101104.
49. Wu, X. Q., Duan, Z. P., Song, H. W., Wei, Y. P., Wang, X., & Huang, C. G. (2011). Shock pressure induced by glass-confined laser shock peening: experiments, modeling and simulation. *Journal of Applied Physics, 110*, 053112.
50. Thord, T., Franz-Josef, K., & Aravinda, K. (2003). Temperatures, pressures and stresses during laser shock processing. *Optics and Lasers in Engineering, 39*(1), 51–71.
51. Ruschau, J. J., John, R., Thompson, S. R., & Nicholas, T. (1999). Fatigue crack nucleation and growth rate behaviour of laser shock peened titanium. *International Journal of Fatigue, 21*, S199–S209.
52. Mannava, S., McDaniel, A. E., & Cowie, W. D. (1996). Laser shock peened rotor components for turbomachinery. US Patent 5,492,447, 20 Feb 1996, General Electric Company (Cincinnati, OH).
53. Mannava, S., McDaniel, A. E., Cowie, W. D., Halila, H., Rhoda, J E., & Gutknecht, J E. (1997). US Patent 5,591,009, 7 Jan 1997, General Electric Company (Cincinnati, OH).
54. Ferrigno, S. J., McAllister, K. G., & Mannava, S. (2001). US Patent 6,200,689, 13 Mar 2001, General Electric Company (Cincinnati, OH).
55. Casarcia, D. A., Cowie, W. D., & Mannava, S. (1996). US Patent 5,584,586, 17 Dec 1996, General Electric Company Cincinnati (OH).
56. Brown, A. S. (1998). A shocking way to strengthen metal. In: *Aerospace America* (p. 21–23).
57. Sokol, D. W., Clauer, A. H., & Ravindranath, R. (2004). Applications of laser peening to titanium alloys. *The ASME/JSME 2004 pressure vessels and piping division Conference*, San Diego CA, 25–29 July 2004.
58. http://www.lspt.com/FatRevChart.pdf.
59. Rankin, J., Hackel, L., & Harrison, J. (2010). Effect of laser peening on fatigue life in an arrestment hook shank application for naval aircraft. *The 2nd international laser peening conference*, San Francisco CA, 19–22 April 2010.
60. Sano, Y. J., Mukai, N., Chida, I., Uehara, T., & Yoda, M. (2010). Applications of laser peening without protective coating to enhance structural integrity of metallic components. *The 2nd international laser peening conference*, San Francisco CA, 19–22 April 2010.
61. Sano, Y. J., Obata, M., Kubo, T., Mukai, N., Yoda, M., Masaki, K., et al. (2006). Retardation of crack initiation and growth in austenitic stainless steels by laser peening without protective coating. *Materials Science and Engineering A, 417*(1–2), 334–340.

# Chapter 2
# Surface Integrity of LY2 Al Alloy Subjected to Laser Shock Processing

**Abstract** This chapter presents a comprehensive literature review of the physical and mechanical mechanisms of LSP on metallic materials, which have been investigated in the past 30 years. In particular, more attentions have been focused on the effects of LSP on mechanical properties of LY2 Al alloys are also highlighted.

## 2.1 Introduction

Since laser shock processing (LSP) was invented in the early 1960s, lots of studies mainly focused on the physics process, the physical mechanisms, the plasma to achieve high pulse pressures [1] and the development of mathematics models to characterize LSP processes [2].

Since 1990, many researchers [3–8] have further developed and enriched this technique by investigating the effects of laser spot shape, characteristics of shock waves and their propagation on the induced mechanical properties. More attention on laser shock processing was paid to some important processing parameters related to LSP conditions, such as laser processing parameter, confined layer and absorbing coatings, which can significantly affect the mechanical properties of the alloys and metallic materials.

This chapter presents laser systems of LSP, particularly addressing its physical mechanisms and effect on the surface integrity of Al alloy. Emphasis is put on LSP processing parameters including laser power density, laser spot, as well as LSP impact number and the coverage ratio of LSP impacts on the nano-hardness, residual stress and surface topography of LY2 Al alloy.

Y. Zhang et al., *Laser Shock Processing of FCC Metals*,
Springer Series in Materials Science 179, DOI: 10.1007/978-3-642-35674-2_2,

## 2.2 Laser Systems for Laser Shock Processing

In order to meet the LSP process requirements, laser requires an average power level from several hundred watts to kilowatts, a pulse energy with several ten Joule and a pulse duration with 8–30 ns. Repetition rate of the laser system and a reasonable laser wavelength are also important parameters for LSP to achieve effective strengthening effect for alloys and metal components.

Neodymium-doped glass (Nd: glass) laser was firstly developed in 1974 at Battelle–Columbus Laboratories Ohio. The repetition rate of this laser with wavelength of 45 m long and with pulse power of 500 J was extremely low, and one cycle can be obtained every 8 min. In 1992, Battelle–Columbus Laboratories invented a $\sim 1.2 \times 1.8$ m Nd: glass laser system with a repetition rate of 1 Hz and a pulse energy of around 100 J based on the above-mentioned technology [9].

Lawrence Livermore National Laboratory (LLNL) has developed high-power Nd: glass laser systems for LSP applications over the past 30 years [10]. These laser systems can deliver average pulse energy of 25–100 J and repetition rates of up to 10 Hz.

The common laser system used for laser shock processing is a high-energy, the wavelength of pulsed Nd: glass laser system is about 1,054 nm high. Suitable laser peening systems can produce short laser pulses, from about 8–30 ns in pulse duration, where the beam energy is in the range of 10–50 J. Although laser systems with lower energy, small spot size or ultra-short pulse durations have been considered, these systems typically do not provide enough power to create the desired effect.

Either circular or square-shaped laser spots can be used depending on the configuration of the laser system and optical components. The laser spot size may be adjusted with lenses, but it is typically an area in the range of 3–28 $mm^2$ that is applied to the part. The LSP parameters are typically selected to achieve a power density or laser irradiance of 5–10 GW $cm^{-2}$. A two-beam laser system may be used advantageously to impact the opposite sides of thin parts, such as the edges of turbine engine airfoils. Laser shock processing one side of a thin blade completely and then the opposite side can result in an excessive distortion. However, this can be avoided by using an alternating scheme in which a laser beam is applied first to one side of the blade and then a second beam is applied to the opposite side. Figure 2.1 shows a two-beam laser system that operates at 50 J per beam.

In the early 1990s, laser systems were not commercially available with the power or laser repetition rate to make laser peening practical. Typically, laser fluence in the range of 100–200 J $cm^{-2}$ is required to create the desired effect in most materials. Early laboratory-based systems had a repetition rate of only one pulse every 8 min. As a consequence, the coverage rate for laser-peening parts was far too slow. The first generation of laser-peening systems was designed with a series of laser amplification stages (all using flash lamps) and an advanced electronic control system. This boosted the laser beam from an oscillator to produce two beams having beam energy up to 50 J per beam and a laser repetition rate of one pulse every 8 s (0.125 Hz).

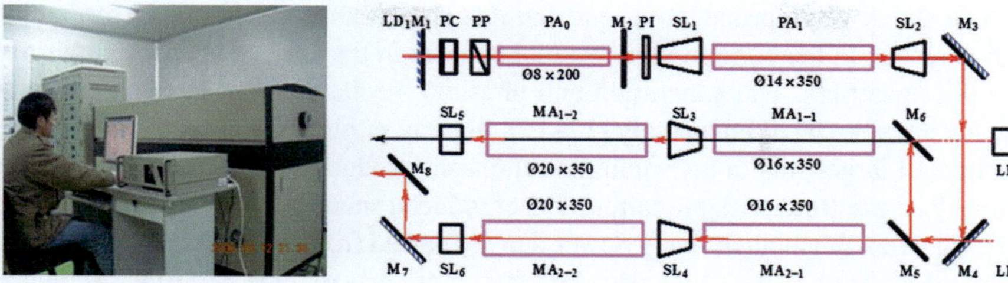


**Fig. 2.1** High-power and repetition-rate Nd: glass pulse laser shock equipment manufactured by Jiangsu University and the light-road layout of laser head. *LD1* ~ *LD3* Semiconductor laser; *M1* 0° mirror; *M2* output mirror, *M3, M4, M6* ~ *M8* 45° mirror; *M5* 45° half-mirror; *PC* KD*P pockels cell; *PP* polarizer; *PI* isolator; *SL1* ~ *SL6* laser bean expander; *PAm, MAm-n* Nd: glass rod

To create the first practical industrial LSP systems, the life and reliability of the laser system components had to increase. The electronics for the pulse-forming networks were redesigned to produce a laser repetition rate of one pulse every 4 s (0.25 Hz) and assembled into modules for easy replacement in case of failure. The materials used for the flash lamp reflectors and the procedures used to produce the laser rods were optimized to boost efficiency and to extend component life. Under the support of the US Air Force ManTech program, LSP Technologies refined LSP technology by building a system with industrially hardened environmental controls and many quick-change features for parts requiring repair or routine maintenance, and boosting the laser repetition rate to one pulse every 0.8 s (1.25 Hz) [11].

Each increase in repetition rate required further improvements to the optical system components to avoid damage during operation. Even higher laser repetition rates for LSP systems are possible (3–5 Hz), even at lower energy levels and using smaller spots. However, laser repetition rate cannot the complete answer to boosting the coverage rate. One must also consider the requirements for applying and replenishing the opaque confined layer and handling the parts. According to these requirements and the laser system, practical LSP coverage rates can be typically 10–20 $cm^2\ min^{-1}$.

The latest laser-peening systems incorporate improved diagnostics to monitor and control the beam's spatial profile (to assess the energy distribution within the laser spot) and temporal profile (to measure the uniformity of the laser pulse timing). These systems also include electronic features for optimizing the shape of the laser pulse to create a uniform shock wave at the surface of the part. This provides a more uniform compressive residual stress distribution in the part being treated.

## 2.3 Generation of Laser Shock Wave

Shock waves can be generated when a laser pulse with high enough power density hits a metal surface. In this way, Q-switched lasers can generate GW laser beams of 1–100 + J pulses less than 100 ns in the action time.

The shock wave propagating into the metal can cause the plastic deformation, which results in the compressive residual stresses in the surface layer of alloys and metallic materials. The generation and physical mechanisms of laser shock wave has been investigated intensively [12–17]. In early published experiments [18, 19], the treated target was in a vacuum and the laser produced a plasma that expanded freely. The resulting peak pressures of laser induced shock wave ranged from 1 up to 10 GPa when the incident laser power density varied from about 0.1–$10^6$ GW/$cm^2$. The time duration of the shock wave pressure pulse was roughly 3–4 times as long as the laser pulse duration in air, but the time duration was only equal to the laser pulse duration in the vacuum because of the rapid adiabatic cooling of the laser generated plasma [20].

The transparent confined layer has been used to increase the intensity of laser shock wave propagating into the metal by up to two orders of magnitude by comparison with plasmas generated in a vacuum [21, 22]. The increase in the intensity of laser shock wave is achieved because the transparent confined layer prevents the generated plasma from expanding rapidly away from the surface. This results in more of the laser energy being delivered into the material as a shock wave [23].

For a Nd: glass laser with a wavelength of 1,054 nm, a transparent confined layer of water, quartz, or glass can increase the intensity of laser shock wave. These transparent confined layers are also known as overlays. The absorbing layer has also been found to increase the intensity of laser shock wave and to protect the metal's surface from laser ablation. Metallic absorbing layers have been found to be beneficial, if not necessary, to protect the component's surface [12]. Among the absorbing layers, commercially available Al foil has been found to be the most practical and effective, as compared to other absorbing layers.

When a laser pulse with enough intensity hits a metal target with an absorbing layer, the absorbing layer absorbs the laser energy, and vaporizes to form a plasma. Because of the ultra-short response times, the diffusion of thermal energy away from the interaction zone is limited to a couple of microns between the confined layer and the top surface of metal target and is preferred to be less than the thickness of the absorbing layer to maintain protection. This is important for alloys and metallic materials because laser ablation will reduce their fatigue lives [24].

The plasma keeps strongly absorbing the laser energy, and expanses in the confined region between the metal target and the transparent confined layer, and then creates a high amplitude, short duration pressure pulse. The shock wave induced by a portion of this energy propagates into the surface layer of alloys and metallic materials. When the peak pressure of the shock wave exceeds the dynamic yield strength of alloys or metallic materials, plastic deformation occurs, resulting in the improvement of the near-surface microstructure and mechanical properties consequently [25].

Some important factors need be taken into accounts when selecting the laser system for laser shock processing, such as cost, efficiency, reliability, and part replacement. Systems suitable for laser shock processing should have an energy output in the range of 10–500 J/pulse with pulse durations of less than 100 ns.

At present, the typical technology that can realize this type of energy output with an acceptable pulse length is a Q-switched laser system based on a Nd: glass or Nd: YAG crystal lasing rod which operates in the near infrared at a wavelength of 1,064 or 1054 nm and pulse duration of 10–100 ns.

## 2.4 Laser Shock Processing Equipment in this Book

In this book, we use two types of laser in the LSP experiment. One is a high-power and repetition-rate Nd: glass pulse laser developed by Laser Technology Institute and the University of Science and Technology of China in 2004, and the laser equipment and the light-road layout of laser head was shown in Fig. 2.1. The shock wave profile of a single laser pulse, and the output luminous field of a single laser pulse were shown in Figs. 2.2 and 2.3. This laser equipment can meet the demands of a single LSP impact and multiple LSP impacts, and its processing parameters were presented in Table 2.1.

The other is Nd: YAG laser manufactured by France Thales Company at Jiangsu University in China. Compared to the above laser equipment, this laser head has some advantages: (1) The Thales Company uses the unique design technology of Nd: YAG ceramic to ensure high stability; (2) Its tight structure and small size bring convenience when it is used; (3) It has a high repetition rate with 5 Hz to make the demand of industrial application. Structure layout of high-power Nd: YAG laser shock equipment manufactured by Thales Company was shown in Fig. 2.4.

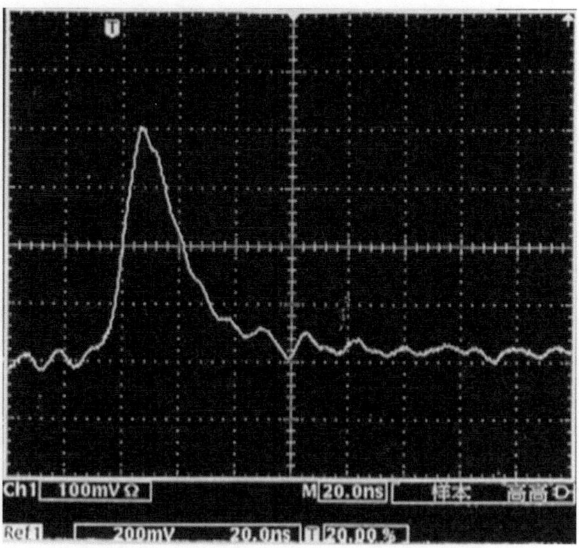


**Fig. 2.2** Laser shock wave profile of a single pulse

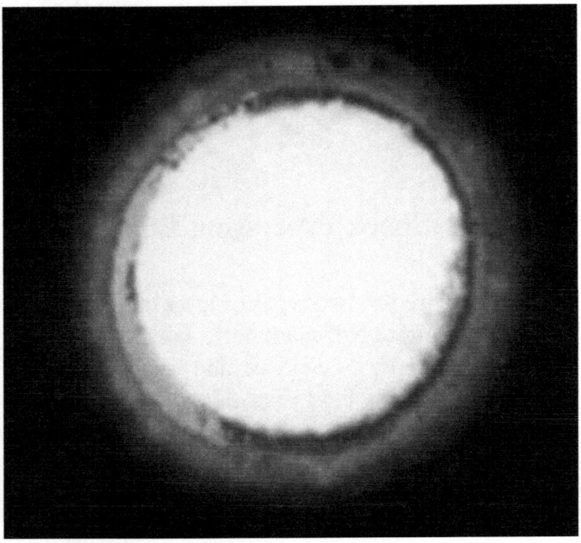

**Fig. 2.3** Output luminous field of a single pulse

**Table 2.1** Processing parameters of Nd: glass laser head

| Type | Value |
|---|---|
| Beam div. of output (mrad) | ≤0.5 |
| Spot diameter (mm) | 1–10 |
| Pulse energy (J) | 10–50 |
| Repetition-rate (Hz) | 1 |
| Laser wavelength (nm) | 1,054 |
| Export stability | ≤±5 % |
| Beam profile | Semi-Gauss |

The system is composed of the following five parts: (1) A computer that controls the power supplies of the laser through the control unit and the ISEO delay generator; (2) The ISEO: 20 channels programmable delay generator; (3) 5 power supplies with a Control Unit for the Laser GAIA—R version; (4) One cooling unit; (5) The laser head which holds all the optical components. The light-road layout was shown in Fig. 2.5.

The output luminous field of a single laser pulse was shown in Fig. 2.6. This laser equipment can meet the demands of a single LSP impact and multiple LSP impacts, and its processing parameters were presented in Table 2.2.

During LSP, the effect of LSP depends on the processing position and the control of shock trace. To realize the accuracy and real-time control on the shock trace and the position between the workpiece surface and laser beam, the self-developed five-axis NC worktable with five degrees of freedom was used in the LSP experiments. The picture of 5-axis NC Worktable was shown in Fig. 2.7.

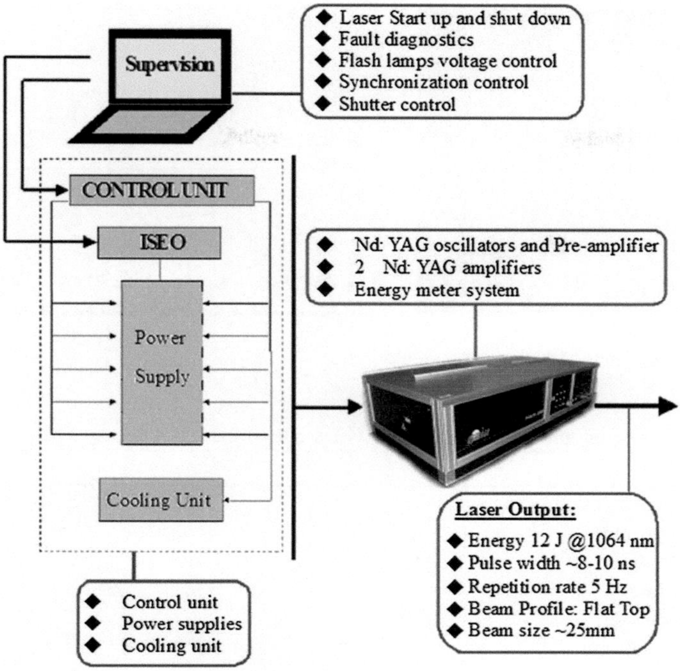


**Fig. 2.4** Structure layout of high-power Nd: YAG laser shock equipment manufactured by Thales company

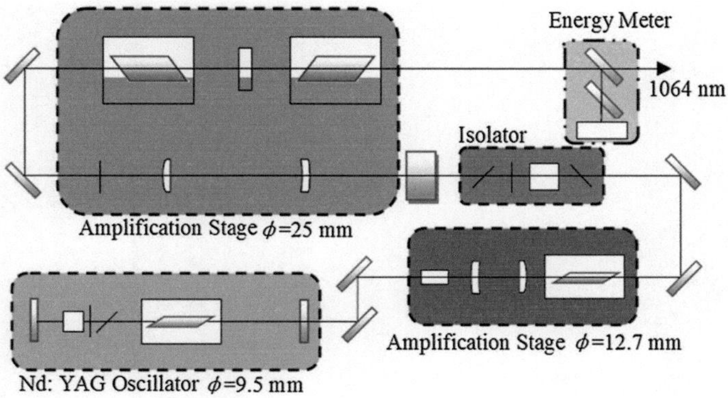


**Fig. 2.5** Light-road layout of high-power Nd: YAG laser head

## 2.5 Nano-Hardness, Micro-Hardness and Residual Stress

After the development of high power laser, the laser surface modification of engineering materials draws a great deal of attention, which in turn causes rapid growth of fields of laser applications such as laser shock processing. The

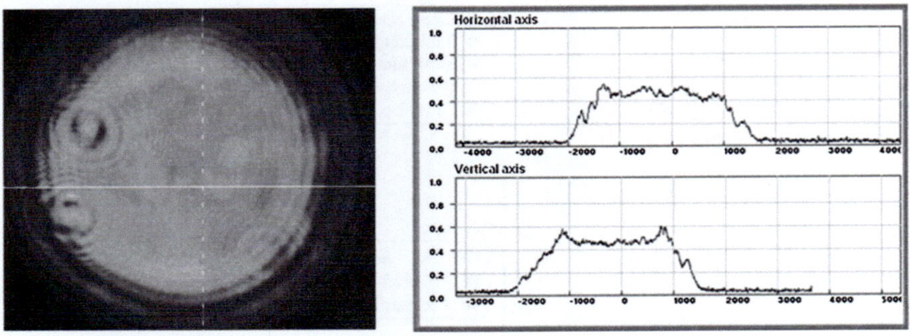


**Fig. 2.6** Output luminous field of a single pulse

**Table 2.2** Processing parameters of Nd: YAG laser head manufactured by Thales company

| Type | Value |
|---|---|
| Beam divergence of output (mm. mrad) | ≤2 |
| Spot diameter (mm) | 2–10 |
| Pulse energy (J) | 3–12 |
| Pulse width (ns) | 8–15 ns |
| Repetition-rate (Hz) | 5 |
| Laser wavelength (nm) | 1,064 |
| Beam profile | Top hat |
| Pulse to pulse energy stability (% rms) | <1.5 % |

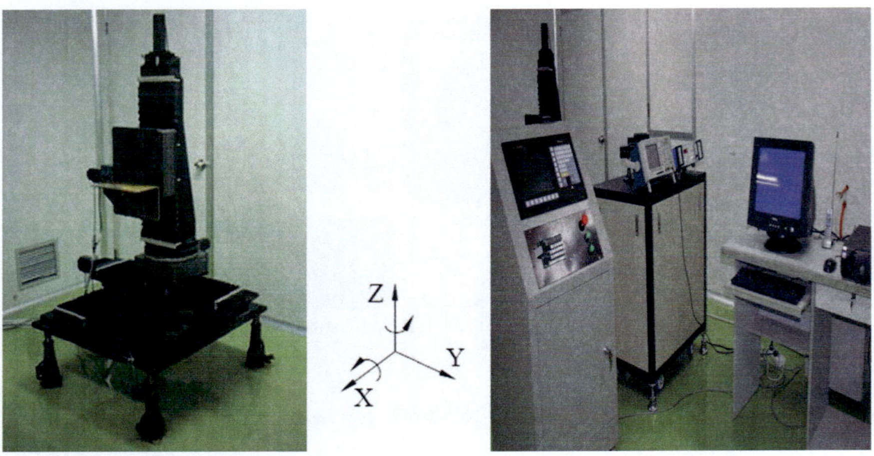


**Fig. 2.7** Self-developed five-axis NC worktable with five degrees of freedom

magnitude of surface residual stress achieved by LSP is usually comparable to that achieved by shot peening, but the depth of compressive residual stress from LSP is significantly greater. Hence, many of the proposed applications of LSP are aimed to improve the fatigue life and strength of the material through surface modification [26, 27].

Al alloy is widely used in aerospace and automotive industries due to its light weight and moderate strength. Considerable researches have been carried out to examine the effects of LSP on mechanical properties and fatigue lives of Al alloys [28–30]. Most of the above researches showed that the mechanical properties and fatigue lives were improved significantly for Al alloys due to the compressive residual stresses in the sample surface and in-depth direction after LSP. However, few studies have been focused on the effects of LSP on the nano-hardness and elastic modulus of Al alloy at the nanoscale. To the treated Al alloy by LSP, the reaction of the laser shock wave and the metal target will be generated near the target surface, leading to the generation of the dislocation and the microstructural deformation near the surface [31], and the increase in hardness was reported to be caused by an increment in the dislocation density with increasing laser impact repetitions [26]. The improvements of hardness and elastic modulus are correlated with the generation of the dislocation and the microstructural deformation near the surface [32] during LSP process, but no fundamental understanding in the LSP literature has been noted on the interaction of the microstructure with laser induced shock waves and the resulting changes in the microstructure. The majority of the reports on microstructural changes have been qualitative with few quantitative details such as dislocation density [33, 34]. The lack of a fundamental understanding on the strengthening mechanism of LSP on Al alloy by microstructure was most clearly reported. The effects of LSP on the nano-hardness and elastic modulus of Fe–Ni alloy were investigated by the nanoindentation techniques, but the mechanisms underlying the improved nano-hardness, elastic modulus and surface residual stress were only preliminarily addressed by evaluating the surface morphologies before and after LSP [35].

With the above background in mind, the purpose of this section was to examine nano-hardness and elastic modulus of LY2 Al alloy after LSP. In addition, the TEM analyses on the microstructures in the different regions after LSP were carried out, and the enhancement mechanism of LSP on nano-hardness and elastic modulus of LY2 Al alloy was also investigated and discussed. These topics discussed in this section will provide some important insights for researchers in the field of surface modification.

### 2.5.1 Experimental Material and Parameters

Some samples manufactured by LY2 Al alloy were cut into a rectangular shape with dimensions of $20 \times 20 \times 2\ mm^3$. The chemical composition and mechanical properties of LY2 Al alloy were shown in Tables 2.3 and 2.4, respectively. Prior to the laser shock, the samples were polished with SiC paper with different grades of roughness (from 500–2,400), followed by cleaning in deionized water. Ultrasound

**Table 2.3** Chemical composition of LY2 Al alloy

| Composition | Cu | Mg | Mn | Si | Fe | Ti | Be | Cr | Zn | Al |
|---|---|---|---|---|---|---|---|---|---|---|
| Percent (wt %) | 2.6–3.2 | 2.0–2.4 | 0.45–0.7 | 0.3 | 0.3 | 0.15 | 0.05 | 0.05 | 0.05 | other |

**Table 2.4** Mechanical properties of LY2 Al alloy

| Type | Value |
|---|---|
| Specific gravity, d, g/cm$^3$ | 2.8 |
| Tensile strength, $\delta$b, kgf/mm$^2$ | 470 |
| Elongation, $\delta$, % | 14 |
| Vickers-hardness, HV | 120 |

in ethanol was used to degrease the sample surface and LSP experiments were conducted shortly after preparation.

During LSP, the shockwaves were induced by a Q-switched Nd: glass repetition-rate laser with a wavelength of 1,054 nm, a pulse of about 20 ns, a diameter of 3 mm, and the laser energy was around 5 J, namely, laser power density is 9.7 GW/cm$^2$. A water curtain with a thickness of about 2 mm was used as a transparent confining layer and the professional Al foil with a thickness of 50 μm was used as an absorbing layer to protect the sample surface from thermal effect.

Transmission electron microscope (TEM) analyses on the microstructures of the different regions were performed on an H-800 TEM (*Hitachi Co., Japan*) with an accelerating voltage of 200 kV.

### *2.5.2 Measurement of Elastic Modulus and Hardness*

Thin film mechanical properties (elastic modulus and yield strength) can be measured by tensile testing of freestanding films [36] and by the microbeam cantilever deflection technique [37], but the easiest way is by means of nanoindentation.

Here, no special sample preparation is required and tests can be performed quickly and inexpensively. Nanoindentation is similar to conventional hardness tests, while it is performed on a much smaller scale using special equipment. The force required to press a sharp diamond indenter into tested material is recorded as a function of indentation depth.

Both elastic modulus and hardness can be readily extracted directly from the nanoindentation curve. Since the depth resolution is on the order of nanometers, it is possible to indent even very thin (100 nm) films. A typical load–displacement curve of a 1 μm thick Cu film is shown in Fig. 2.8.

Elastic modulus is determined based on the knowledge of the tip shape function, and the load–displacement curve (load P and displacement h) [38].

The nano-hardness $H$ and elastic modulus $E_r$ were defined as follows:

$$H = \frac{P_{\max}}{A} \tag{2.1}$$

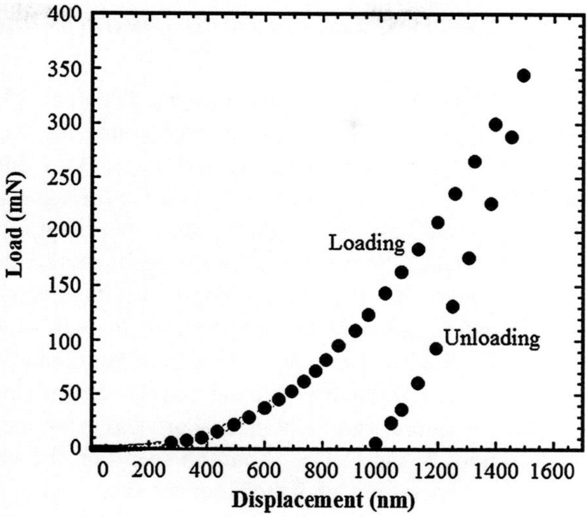


**Fig. 2.8** Load–displacement curve for a 1 μm thick Cu film

$$E_r = \frac{\sqrt{\pi}}{2\beta}\frac{S}{\sqrt{A}} \tag{2.2}$$

here $P_{\max}$ is the maximum load, $A$ is the projected contact area, $S$ is the banner slope of unloading curve and $\beta$ is the contact compliance between the indenter and the sample, which is equal to the tangent to the force–displacement curve during unloading after correction for frame compliance.

The symbols $A$ and $h_c$ in the Eqs. (2.1) and (2.2) are calculated as:

$$A = 24.56h_c^2 + \sum_{i=0}^{7} C\beta h_c^{1/2^i} \tag{2.3}$$

$$h_c = h - \beta P_{\max}/S \tag{2.4}$$

where $h_c$ is the contact depth and $h$ is the indention depth, as shown in Fig. 2.9.

The other experimental conditions of nano-indentation are: (1) Depth limit 3,000 nm; (2) Strain rate target 0.05 $s^{-1}$; (3) Maximum calculation depth 2,500 nm and (4) Minimum calculation depth 1,000 nm.

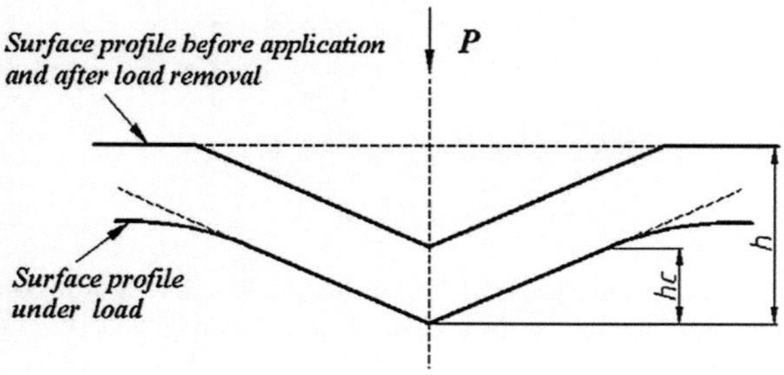


**Fig. 2.9** Definition of depth hc and h in the indentation test

### 2.5.3 Effects of LSP on Nano-Hardness and Elastic Modulus

Figure 2.10a, b, c showed the measured load–displacement curves for the non-shocked region, the laser-affected region and the laser-shocked region at a maximum load of 960 μN, respectively. $P_{\max}$, $A$, $S$ and $h_c$ were calculated during unloading by using Eqs. (2.1–2.4), and the measurements were repeated five times for each condition, an average value was determined on the basis of the five measured data. The corresponding experimental data were shown in Table 2.4. It can be found that the contact depths of the indenter were 247.58, 196.07 and 150.51 nm in the non-shocked region, the laser-affected region and the laser-shocked region, respectively. The maximum contact depth of the indenter at peak load decreased gradually from the non-shocked region to the laser-shocked region.

The measured values of nano-hardness near the surface in different regions were shown in Fig. 2.11a. It can be seen that the values of nano-hardness in the laser-shocked region and the laser-affected region were improved obviously in comparison with the corresponding values in the non-shocked region. The experimental data about nano-hardness $H$ were also shown in Table 2.5 in detail. Firstly, it can be seen from Table 2.4 that in the non-shocked region, i.e., the untreated material, the nano-hardness kept a constant value of 0.535 GPa;

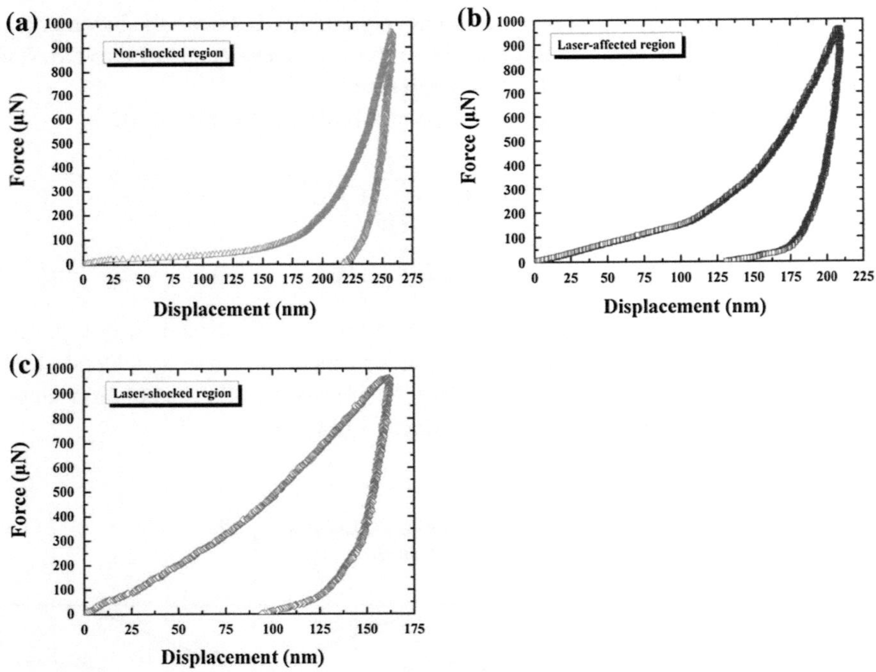


**Fig. 2.10** The measured load-displacement curve at a maximum load of about 960 μN in the indentation test. **a** In the non-shocked region, **b** in the laser-affected region, **c** in the laser-shocked region

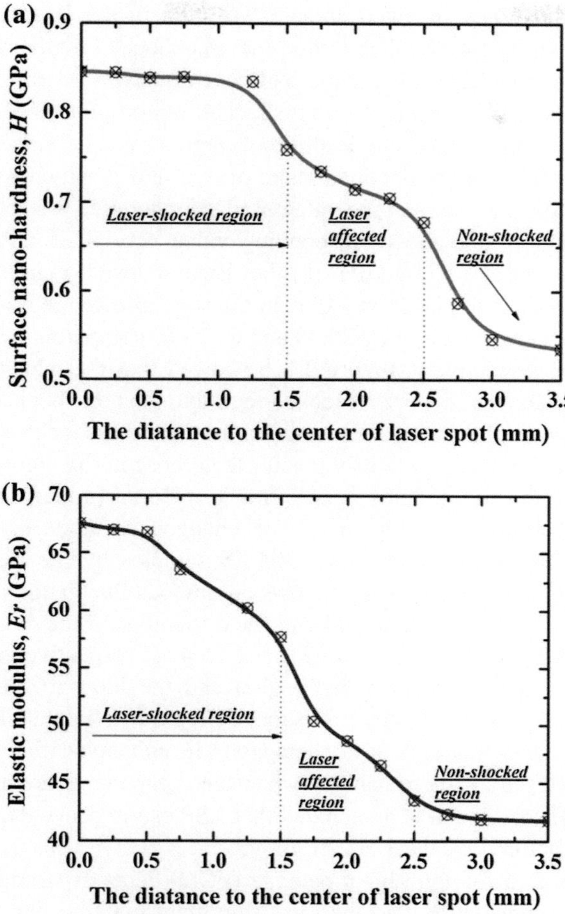


**Fig. 2.11** The measured values near the surface in different laser treated regions in the indentation test. **a** nano-hardness, **b** elastic modulus

**Table 2.5** The values of nano-hardness and elastic modulus in different regions

| | Non-shocked region | Laser-affected region | Laser-shocked region |
|---|---|---|---|
| Maximum load, $P_{max}$/μN | 960 | 960 | 960 |
| Contact depth, $h_c$/nm | 247.58 | 196.07 | 150.51 |
| Contact area, $A$/nm$^2$ | 1,142,091 | 1,381,584 | 1,783,028 |
| Slope, $S$/μN·nm$^{-1}$ | 47.23 | 61.74 | 72.95 |
| Nano-hardness, $H$/GPa | 0.535 | 0.718 | 0.846 |
| Elastic modulus, $E_r$/GPa | 41.77 | 50.33 | 67.56 |

Secondly, nano-hardness reached about 0.718 GPa in the laser-affected region, which was greater by 34.2 % than that in the non-shocked region; Lastly, when it got into the laser-shocked region, nano-hardness value was 0.846 GPa, which was greater by 58.13 % than that in the non-shocked region after LSP treatment.

The values of elastic modulus in different regions were obtained according to Eq. (2.2) on the basis of the obtained value of $A$, $S$ and $\beta$, which were also shown in Table 2.4. It can be seen that the value of elastic modulus was 41.77 GPa in the non-shocked region, and the corresponding value was 50.33 GPa in the laser-affected region, while 67.56 GPa in the laser-shocked region, respectively. Obviously, the values of elastic modulus in the laser-affected region and the laser-shocked region were increased by 20.49 and 61.74 % compared to that in the non-shocked region, respectively. Figure 2.11b showed that the magnitude of elastic modulus decreased from the laser-shocked region to the non-shocked region, and it was clear that the values of elastic modulus in the laser-shocked region and the laser-affected region were obviously greater than those in the non-shocked region.

Although the elastic modulus is an intrinsic material property and fundamentally related to atomic bonding, it can be changed by some surface treatment technologies. The elastic modulus of 304 SS samples by the high-energy shot peening and ultrasonic shot peening shows an obvious fluctuation along the depth profile [39]. The elastic constant C44 and shear modulus of the Al alloy 2024-T62 subjected to LSP decrease by up to 29.0 and 25.4 %, respectively. The values of elastic modulus in the laser-shocked region and the laser-affected region were clearly greater than those in the non-shocked region. It is well known that the increment of elastic modulus is advantageous in enhancing the stiffness of the parts, which determines their stability when the components are applied in environment [40]. Hence, it can be concluded that LSP can improve the stiffness of the samples manufactured by the LY2 Al alloy.

The resistance of foreign object damage (FOD) depends strongly on the stiffness of the component [41] and the most important material parameter affecting FOD was identified as hardness of the surface material [42]. From the above experimental data, it can be seen that the values of nano-hardness in the laser-shocked region and the laser-affected region were greater than those in the non-shocked region. As a consequence of the lower contact depth in the laser-shocked region compared to that in the non-shocked region, the values of nano-hardness in the laser-shocked region were greater from another point of view. When shock wave pressure was higher than approximately 5 GPa, LSP can produce the increments of surface hardness over the entire laser irradiated area, such as the hardness of underaged 2024-T351 [43] and 6061-T6 [44]. The above research results were also in good agreement with those obtained by Montross et al. [45]. Therefore, LSP can improve the nano-hardness and elastic modulus of sample surface layer, which was favorable for improving FOD resistance of the components manufactured by LY2 Al alloy.

### 2.5.4 Enhancement Mechanism of LSP on Hardness and Elastic Modulus

Figure 2.12a, b showed the TEM micrographs of sample surface with the laser power density of 9.7 GW/cm$^2$ in the non-shocked region and in the laser-shocked region, respectively. It can be seen from Fig. 2.5a that there were plenty of grains with the average dimensions of 130 $\times$ 130 nm$^2$ and a small amount of dislocations in the non-shocked region. In the laser-shocked region, the grains were clearly refined to be the smaller grains with the average dimensions of 60 $\times$ 60 nm$^2$, and there were plenty of dislocation tangles in the vicinity of grain boundaries. In addition, a small amount of twins were also found (which were demonstrated by the rectangulars A, B and C). Figure 2.12c was the high magnification micrograph of the twins of the rectangular A in Fig. 2.12b.

Micro-mechanism of the metal plastic deformation is a generation and motion of dislocation, and the metal plastic deformation $\varepsilon$ can be expressed by the Orowan's relation [46], i.e.

$$\varepsilon = K\rho\chi b \tag{2.5}$$

where $\rho$ was dislocation density, $K$ was a factor that was concerned of dislocation, $b$ was the scalar value of Burgers vector, and $\chi$ was the average distance of dislocation.

According to the previous work [47], mechanical properties, such as hardness and yield stress, can be expressed in terms of microstructure, such as dislocation

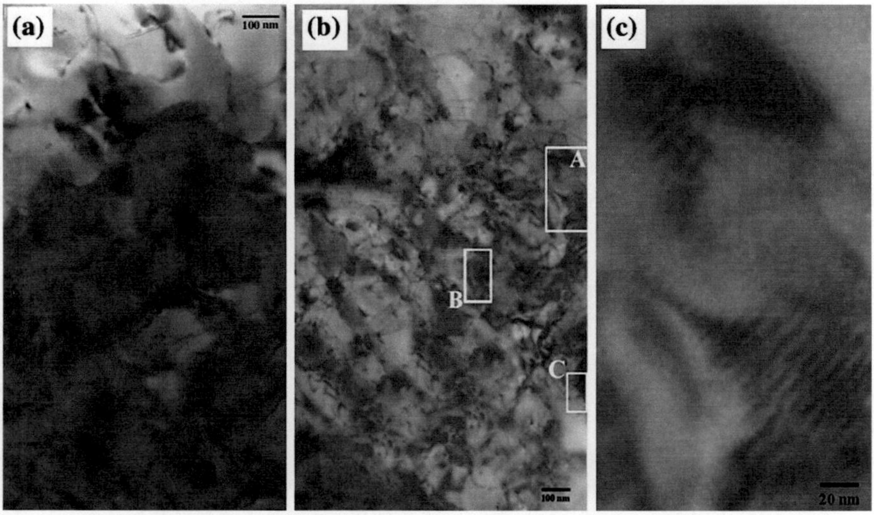


**Fig. 2.12** (**a**) and (**b**) is respectively the TEM micrographs of sample surface with laser power density of 9.7 GW/cm$^2$ in the non-shocked region and the laser-shocked region, and (**c**) was the magnification TEM micrograph of the twins of the *rectangular A* in (**b**)

density and twin spacing. In the present study, the dislocation substructure was the key feature observed in the laser-shocked region of LY2 Al alloy. The measured surface nano-hardness, *H*, can be described by the following equation,

$$H = H_0 + aGb\sqrt{\rho} \tag{2.6}$$

where $H_0$, $a$, $G$, and $b$ are the material's constants. $H_0$ is the surface nano-hardness of an ideal material without any defects, $G$ is the shearing modulus; and $a$ is the constant which has relation to crystal structure. As an approximation, the material's constants $a$, $G$, and $b$ are calculated based upon the results obtained by Ganin et al. [48]. In the present study, LY2 Al alloy was explosively shock loaded at 3.14 GPa and nano-hardness values are taken on sample surface in different regions.

The microstructural changes occurred in the alloy during LSP, and the plastic deformation were a direct consequence of the change of grain size and dislocation density and the rearrangement of dislocations. The reaction between laser shock wave and the metal target will be generated near the target surface after LSP, leading to the increment of the dislocation density and the microstructural deformation near the surface [49]. LSP caused plastic deformation of the sample surface which was recessed by several microns due to this plastic deformation, and high density dislocations and a small amount of twins were observed in the TEM micrographs. Therefore, it was reasonable to conclude that the strengthening of LSP on the mechanical property of LY2 Al alloy was mainly due to the generation of high density dislocations after LSP.

## 2.6 Surface Roughness and Surface Profile

Surface roughness plays a highly important role on fatigue lives of alloys [50, 51]. The previous studies very little involve the fully investigation of surface topography about LSP treatment [52–54]. Zhang et al. [52] investigated the surface qualities of LSP zones to the fatigue life of Al alloy. They found that the fatigue life decreased with the increase of surface roughness. Chu and Rigsbee [47] found that the surface roughness of low carbon steel increased with the increase of laser shock pressure. Rozmus [53] found the surface roughness of Ti6Al4 V Alloy increased from 0.1–0.82 μm after LSP treatment. However, the research of Zhang and Lu [33] indicated that the surface roughness of LY2 Al alloy decreased obviously after LSP treatment. The reason caused the difference of conclusion was that they all did not consider the effect of initial surface topography to the results of LSP treatment. Additionally, the effect of initial surface topography to the residual stress and micro-hardness has also not been discussed.

Different mechanic methods will produce different kinds of surface topography and corresponding typical features. It is important to understand the evolution of these typical features because their existence will acquire different results as treated by LSP. In this study, LY2 Al alloy samples with four different initial surface conditions were prepared. The surface topography was carefully investigated before and after LSP treatments. In addition, the surface residual stress and micro-hardness were also studied. These studies can provide more clues with regarding to fatigue lives of LY2 treated by LSP.

### *2.6.1 Sample Preparation and Surface Topography Measurements*

The samples of LY2 were cut into rectangular plates with dimensions of $10 \times 10$ mm and a thickness of 5 mm, and then they were mounted on a motor controlled by $x - y$ stage with 1 μm of resolution and maximum speed of 20 mm $s^{-1}$. Four different initial surface conditions were conducted as following: 1 burnished with #150 SiC paper, 2 burnished with SiC paper with different grades of roughness (from #150 to #600), 3 burnished with SiC paper with different grades of roughness (from #150 to #1,200), 4. burnished with SiC paper with different grades of roughness (from #150 to #1,200), then polished with silk. Before and after LSP, ultrasound treatment in ethanol was used to clean the surface of all the tested samples.

LSP experiments were conducted shortly after the preparation of LY2 samples. LSP experiments were performed using a Q-switched Nd: YAG laser operating at 5 Hz repetition rate with a wavelength of 1,064 nm, and the full width at half maximum (FWHM) of the pulses were about 10 ns. The laser beam was focused by a focal lens and the spot diameter was about 3 mm when it arrived at the to-be treated samples. A water layer with a thickness of about 1 mm was used as the transparent confining layer and a professional black polyester tape with a thickness of 180 μm was used as the absorbing layer. The laser pulse energy was 6 J. During multiple LSP impacts, the laser beam was perpendicular to the same location on the sample, and the black polyester tape was replaced after each impact.

Surface topography was test by a Veeco Wyko NT 1,100 non-contact optical profiler (NCOP). The Wyko Vision software is the heart of the optical profiler measurement system. It provides powerful measurement and analysis capabilities, with over 200 analysis tools including 2D and 3D plots, histograms, and many application-specific tools and parameters, including surface height statistics such as Rp, Rv, Rt Rq, and Ra [55]. In this section, the tested area was a $240 \times 180$ μm rectangle at the center of laser spot.

### 2.6.2 Surface Topography

Figures 2.13a1, a2, a3 show the 3D surface topography of #150 samples before LSP, by one LSP impact and by three LSP impacts respectively. In Fig. 2.13, #150 and Un represents the sample burnished with #150 SiC paper before LSP, #150 and One LSP represents the sample burnished with #150 SiC paper then subjected to one LSP impact and #150 and Three LSP represents the sample burnished with #150 SiC paper then subjected to three LSP impacts, respectively. The corresponding 2D topography is shown in Fig. 2.14. Compared with #150 and Un (Fig. 2.13a1), #150 and One LSP (Fig. 2.13a2) and #150 and Three LSP (Fig. 2.13a3), it can be seen that the features marked with circle disappear gradually. The characteristic of these gradually disappeared features can be seen in Fig. 2.14. The height-width ratio for these kinds of features is high (more than 2). These features are named sharp protruding here. However, it can be seen that the features marked with parallelogram always exist even after three LSP impacts. The typical characteristic of these features is micro groove.

In order to find out the detailed statistical data of examined surfaces, the number of data points at each surface height is shown Fig. 2.15. The sampling of surface height is dependent on the Wyko Vision software automatically. It can be seen from Fig. 2.15a that the number of data points whose height exceeding 4 μm decreases gradually after LSP. The surface height distribution of #150 and Un scatters from −5–15 μm, and it is narrowed effectively after one LSP (from −5–4 μm).

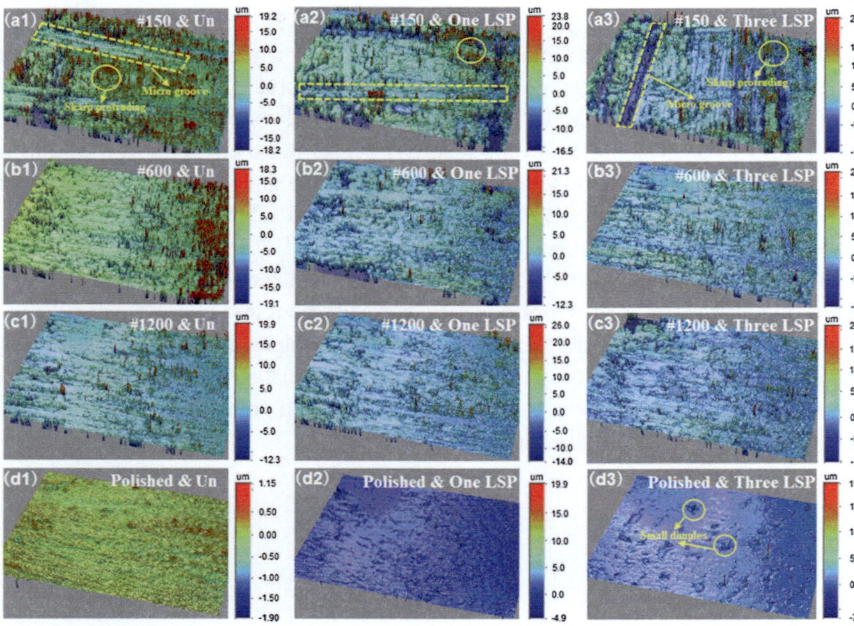


**Fig. 2.13** 3D surface topography of LY2 Al alloy treated by LSP

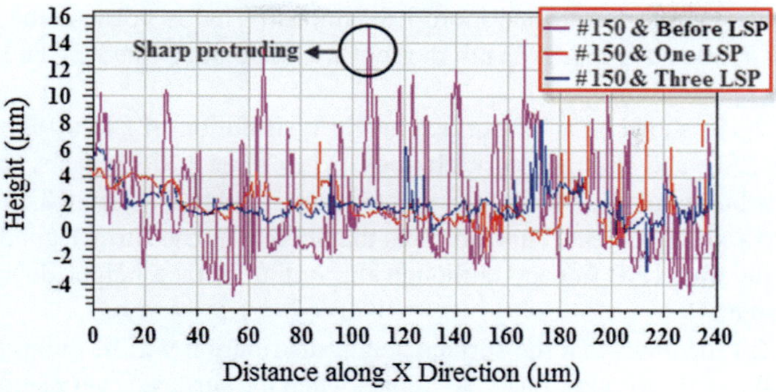


**Fig. 2.14** 2D surface topography of LY2 Al alloy burnished with #150 SiC paper

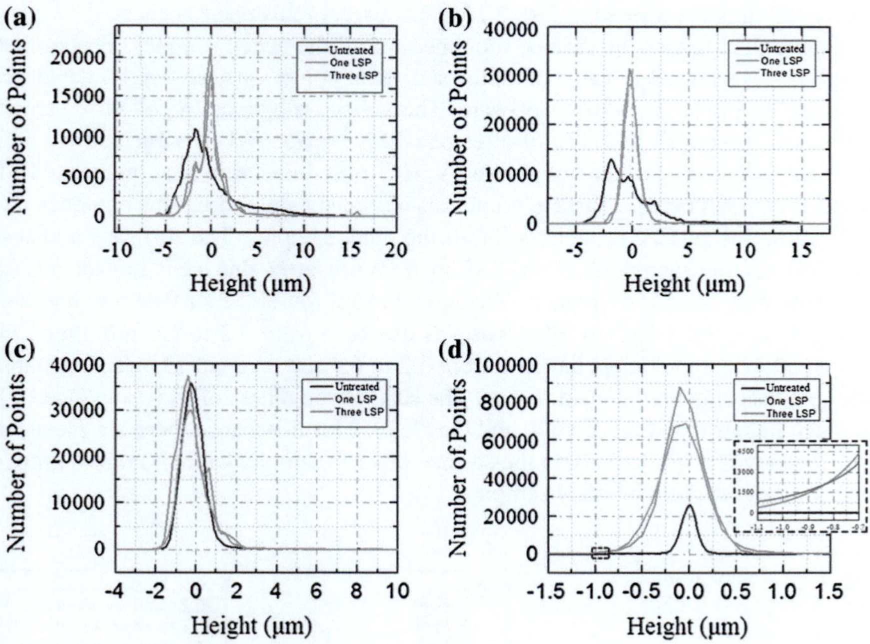


**Fig. 2.15** Histogram showing the number of data points at each surface height. **a** burnished with #150 SiC paper, **b** burnished with #600 SiC paper, **c** burnished with #1,200 SiC paper, **d** polished with silk

When subjected to three LSP impacts, there is hardly change for surface height distribution.

The surface height of #600 and Un (Fig. 2.15b) scatters from −5 to 5 μm, while it scatters from −2.5 to 2.5 μm after one LSP. When subjected to three LSP impacts, it has a little change (from −2.0 to 2.5 μm). The surface height distribution of #150 and One LSP and #600 and Un is almost similar, but the results are different when subjected to one more LSP impact. This is because the dominant features of #150 and One LSP is micro groove while that of One LSP and #600 and Un is sharp protruding.

Figure 2.15c shows that the surface height distribution of #1,200 is widened a little after LSP, and it is not very obvious. After three LSP impacts, the surface height distribution of #1,200 sample scatters from −2.0 to 2.5 μm, which is similar to the #600 and Three LSP. This indicates that their ultimate surface condition may be the same after LSP impacts although the initial surface height distribution is quite different.

Figure 2.15d shows that the surface height distribution will be enlarged rapidly (from −1.0 to 1.0 μm) after one LSP impact when the surface is very smooth (from −0.4 to 0.4 μm). When subjected to three LSP impacts, it seems that there is a little difference compared with one LSP impact. However, the left partial enlarged view shows that the number of points below −0.85 μm increases after three LSP impacts. The small dimples formed in Fig. 2.15d3 can explain this phenomenon.

Figure 2.16 shows the surface roughness of all the tested samples. The surface roughness is an average value of all the statistical points' surface height, which are sampled by the Wyko Vision software. The surface roughness $R_a$ of #150 sample decreases from 2.45 to 1.47 μm after one LSP impact. When subjected to three LSP impacts, the surface roughness is 1.42 μm, indicating that multiple LSP impacts cannot decrease surface roughness obviously any more. This tendency can be seen on the surface roughness for all the other samples. The surface roughness of #600 sample decreases from 1.88 to 0.59 μm after one LSP impact and to 0.58 μm after three LSP impacts. Zhang et al. [52] found the surface roughness of some of their 2024-T72 Al alloy samples decrease from 3.2 to 1.6 μm after LSP treatment, but some others increase from 3.2 to 6.4 μm, even to 12.5. The unstable LSP process may be the main reason. The surface roughness of #150 and One LSP is much higher than that of #600 and One LSP. This is because there are plenty of micro grooves (Fig. 2.13a3) on the surface of #150 sample, while no micro groove exists on the surface of #600 sample.

**Fig. 2.16** Comparison of surface roughness with different surface status

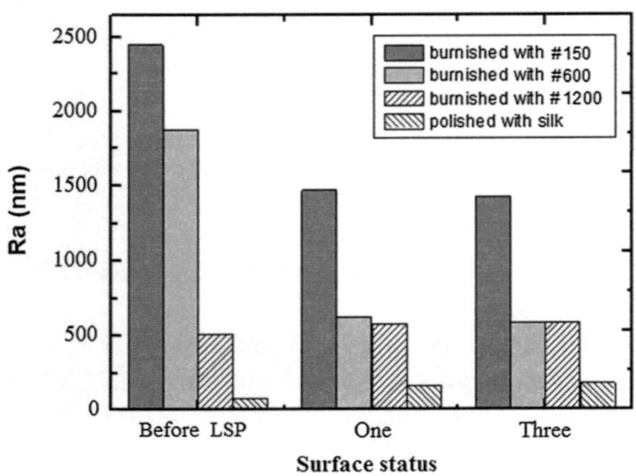

However, the surface roughness of #1,200 sample increases from 0.49 to 0.56 μm after one LSP impact and to 0.58 μm after three LSP impacts. Although the initial surface roughness is quite different, the surface roughness of #600 and Three LSP is equal to that of #1,200 and Three LSP. These imply that there is an ultimate value about surface roughness of LY2 in the present experimental conditions. Similar results could be found in the works of Luong et al. [26]. They found the surface roughness of 7085-T7651 Al alloy increased from the initial 0.444 to 0.518 μm after three LSP impacts. Rozmus-Górnikowska et al. [56] found the roughness of Ti6Al4 V Alloy increased from 0.1 μm before treatment to 0.82 μm after LSP impact. All these results indicate that there is some ultimate value about surface roughness surface roughness for the alloys treated by LSP, but their ultimate values are different from each other. The reason of this phenomenon may be the difference and parameter of LSP treatment mechanical performance of alloys.

For the polished sample, the surface roughness of sample before LSP is double that of one LSP impact, and then a very small increment is obtained when subjected to three LSP impacts. It can be seen that small dimples (Fig. 2.13d3) form when subjected to three LSP impacts, which will certainly enhance the surface roughness. So the surface roughness of polished sample can be continuously increased. Similar result also could be found in the study of Chu et al. [47], in which they found the increase of surface roughness when a polished sample of low carbon steel is subjected to LSP impact. The underlying reason for the increase of surface roughness may be due to the unevenness of absorbing layer. Figure 2.17 shows the schematic principle of this phenomenon. Here, the surface of alloy is assumed to be flat. When laser shock wave propagates into the air between the Al foil and alloy, it will attenuate. Obviously, this will cause the intensity difference of laser shock wave on the surface of alloy at different positions. The intensity difference of laser shock wave will certainly bring the increase of surface roughness.

It may be the fact that after enough number of LSP impacts, when the vanishing of sharp protruding and the production of small dimples caused by the un-uniform of laser intensity reach to a balance, the surface roughness reaches its ultimate value. Here the ultimate value is 0.58 μm.

For the #150 sample, it can also be reasonably concluded that the surface roughness will reach to 0.58 μm when subjected to enough number of LSP impacts at a same point. However, it is not economical in industrial production as to multiple LSP impacts.

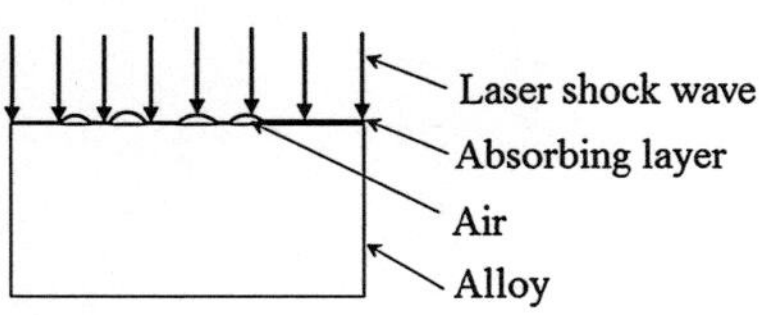


**Fig. 2.17** Schematic illustration of the increase of surface roughness generated by laser shock processing

### 2.6.3 Residual Stress

Because the residual stress is different everywhere along the radial direction [57] within a same laser spot when treated by LSP, all the tested points locate in the center of laser point to make a suitable comparability. Figure 2.18 shows the tested surface residual stress of all the samples. The surface residual stress of #150, #600, #1,200 and polished sample is −58, −133, −146 and −149 MPa respectively when subjected to one LSP impact. This indicates that the increasing surface roughness will decrease the compressive residual stress. When subjected to three LSP impacts, the surface residual stress of all the samples are almost the same (about −150 MPa). It can be concluded that the residual stress tends to be saturated when subjected multiple LSP impacts and the saturated compressive residual stress is independent of surface roughness.

### 2.6.4 Micro-Hardness

The surface micro-hardness is also different everywhere [58] on the scope of same laser spot, which is similar to the distribution of residual stress. In order to ensure good comparability of the results, the testing points of micro-hardness are around the center of laser spot. Figure 2.19 shows the surface micro-hardness of all the tested samples. The micro-hardness of all the samples before LSP is about the same. It is obvious that LSP can improve micro-hardness. After one LSP impact, the micro-hardness decreases with the decrease of surface roughness. The reason may be described as Fig. 2.20: LSP will cause sharp features to vanish, and these features will experience much more severe deformation than the flat surface. This reinforces work-hardening effects and causes the increase of micro-hardness. After three LSP impacts, the micro-hardness of all the tested samples increases slightly. The increase of micro-hardness is due to the increase of dislocation density, this

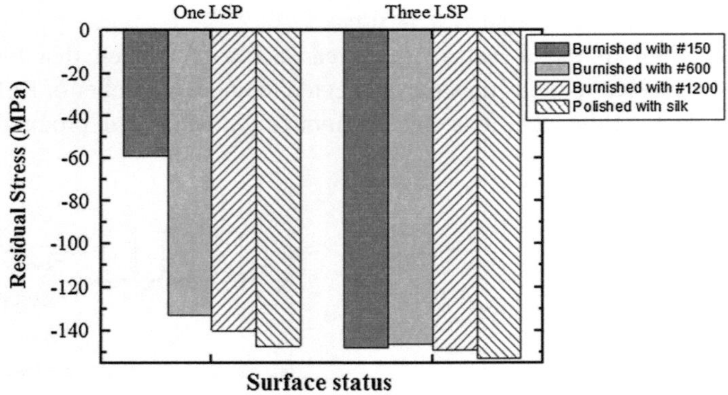


**Fig. 2.18** Comparison of surface residual stress with different surface status

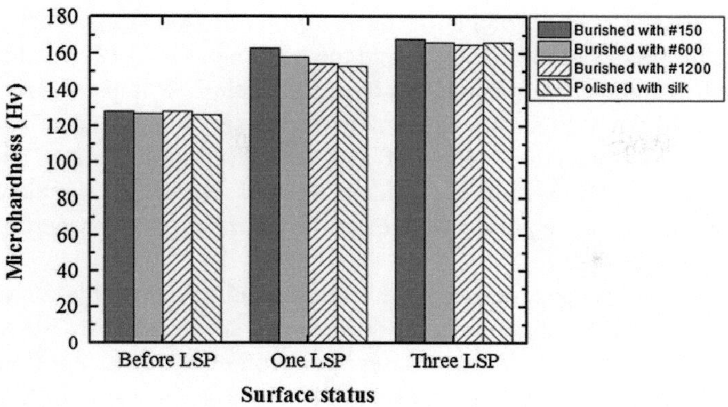


**Fig. 2.19** Comparison of surface micro-hardness with different surface status

**Fig. 2.20** Schematic of plastic deformation with sharp protruding

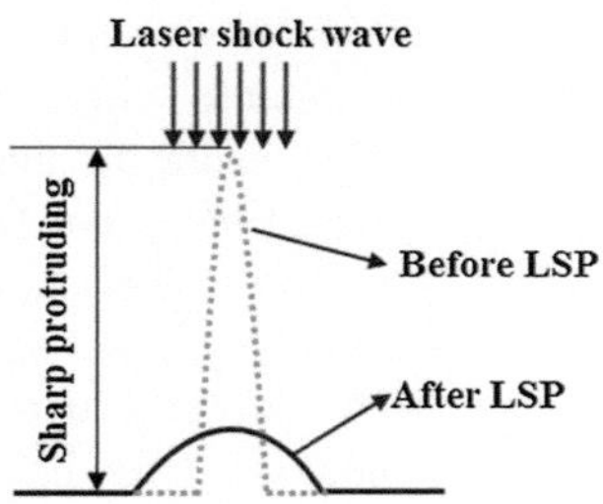


has been proved in the work of Clauer et al. [59]. However, the micro-hardness of all the samples is almost equal after three LSP impacts. This implies that the initial surface topography has no influence on the micro-hardness of LY2 after multiple LSP impacts.

## 2.7 Simulation and Validation of the Residual Stresses Using Laser Elliptical Spot

Among the laser surface treated methods, the laser shock processing (LSP) is a new and promising surface treatment technique to improve the fatigue, corrosion and wear resistance and the mechanical properties of the metals and alloys through generating this shock wave and introducing the deep compressive residual stresses into the materials [60–67]. Hence, many of the proposed applications of LSP are aimed to improve the fatigue life and strength of the material through surface modification [68, 69].

Warren et al. [67] performed finite element method (FEM) simulations of LSP Al(Al) alloy in single and multiple passes by using the developed spatial and temporal shock pressure model. The study results indicated that high residual compressive stress was developed in the deformed region [22], which improved the surface properties of the Al alloy. And the predictions of the FEM analysis [70] agreed well with the experimental data of the multiple LSP impacts. Hu et al. [71] have performed three-dimensional (3D) FEM analysis on the residual stresses due to multiple LSP, and the effects of different overlapping rates and impact sequences on the distribution of residual stresses were also simulated. Some experimental results were presented to examine the mechanical properties of metal samples after LSP [72, 73].

The above studies have focused on the determination and simulation of residual stress and hardness distributions at the surface and subsurface by LSP with round spot. It is well known that different laser spot profiles can induce different shock effects, and the different spot profiles can also be applied according to different requirements during LSP. A laser beam with a round profile was extensively used due to the work matter of the laser, and a laser beam with a square profile allowed dense, uniform packing of the laser spots [71], the distribution of residual stresses by LSP with square spot was investigated in the literatures [74, 75], which was different compared with that with round spot. Clauer et al. [76] focused on the effect of LSP on the fatigue crack growth of pre-existing cracks of the samples with a centered hole, using different laser spot profiles, such as annular spot and round spot. The results showed that the 2024-T3 samples treated with the round laser spot had a fatigue life about 13 times longer than the ones with the annular laser spot. Meanwhile, there is another laser spot profile, i.e., elliptical spot, which is easily modulated and has been applied for long edges of metal parts. However, LSP with elliptical spot is rarely investigated.

The aim of this section was to investigate the effect of laser shock processing with elliptical spot (long-axis length 12 mm and the short-axis length 3 mm) on the residual stress distribution of LY2 Al alloy samples. Residual stress was investigated with different overlapping rates. A FEM model was developed to simulate the distribution of residual stress during the overlapping process by using ABQUAS software. The simulated residual stress field was analyzed and compared with the experimental results. The relations between the magnitude and uniformity of residual stress and the overlapping rate were also addressed. These discussed topics could give some important insights on the surface treatment and life-extension of long edges of metal parts.

### *2.7.1 Sample Preparation and Measurements of Residual Stress*

The LY2 Al sheet was cut into a rectangular shape with $30 \times 15 \times 10$ mm$^3$ (width $\times$ length $\times$ thickness). The grinding and polishing were performed across the sample surfaces, and then the samples were cleaned ultrasonically before the experiments.

In LSP, the shockwaves were induced by a Q-switched repetition-rate laser with a wavelength of 1,054 nm and a pulse of around 20 ns. The water with a thickness of 1–3 mm was used as the transparent confining layer and the 7,075 Al foil with a thickness of 100 um was used as an absorbing layer to protect the blade surface from thermal effect. Laser energy was around 40 J. The deformed surface layer is loaded in compression by the undeformed bulk material as shown in Fig. 2.21a, b.

Five samples by LY2 Al alloy were treated by using the same experiment condition, and the schematic diagram of the shocking path by LSP with elliptical spot was shown in Fig. 2.22.

The residual stresses at the surface of the sample and through the depth direction after LSP were determined by using x-ray diffraction. For the measurement of the residual stress in depth direction, the electropolishing material removal method was used. The experiment was carried out with overlapping LSP. The experiments were repeated five times, and the experimental data are the average value of five samples.

## 2.7.2 Numerical Simulation Procedures

The commercial FEM software ABAQUS [77] was used to predict the residual stress and strain distribution at the surface and subsurface of the metal target during and after LSP. As most of the laser impacts were elliptical, a plane

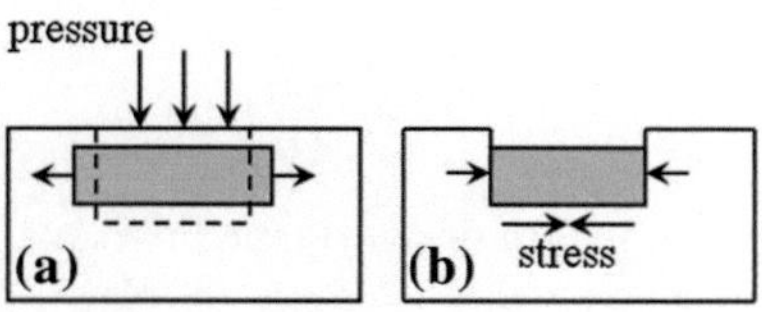


**Fig. 2.21** Residual stresses pattern during (**a**) and after (**b**) the interaction

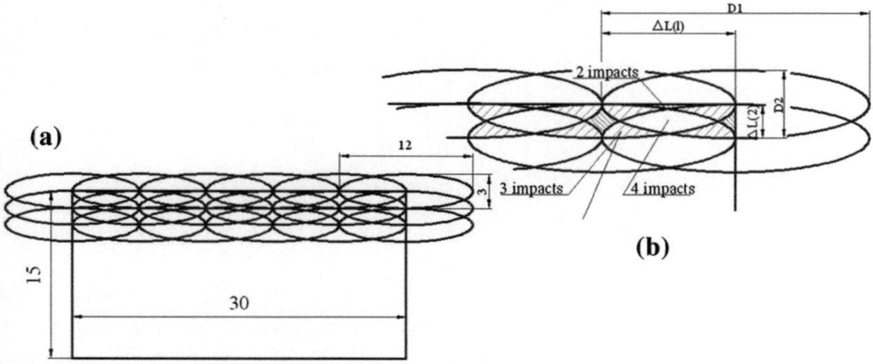


**Fig. 2.22** Schematic diagram of the shocking path by LSP with elliptical spot when the long-axis length is 12 mm and the short-axis length is 3 mm

symmetric problem was chosen in order to reduce the data processing time. C3D8R (continuum, 3 dimensional, eight nodes, and reduced integration) was used in the FEM. The symmetric boundary condition was employed on the XOZ plane of the 3D finite element area. The FEM analysis procedure of LSP should be composed of two distinct parts [78], namely dynamic analysis and static analysis to capture an absolutely stable residual stress field and the elasto-plastic surface deformation.

In the present work, the residual stresses subjected to the single and multiple LSP with elliptical laser spots was considered in the experimental investigation and the FEM analysis. The material properties were listed in Table 2.4. LSP generated strain rate at the surface of the target exceeding $10^7\ s^{-1}$. The elastic limit stress in the direction of the shock wave propagation was defined as the Hugoniot Elastic Limit (HEL) [70]. When the peak pressure in the direction of the wave propagation is greater than HEL, plastic deformation occurs. The relationship between the dynamic yield strength, $\sigma_y^{dyn}$, and HEL under uniaxial strain conditions can be defined as

$$\sigma_y^{dyn} = \mathrm{HEL}\frac{1 - 2v}{1 - v} \tag{2.7}$$

where $v$ is the Poisson's ratio. The plastic strain was assumed to follow the von Mises yielding criterion in which the dynamic yield strength was defined as above.

Overlapping rate $\eta$ in two directions is defined to describe and control the distribution of laser pulses on the impact region, which is expressed as

$$\eta = \frac{\Delta L}{D} \tag{2.8}$$

where $\Delta L$ is the coincidence length of overlapped two successive laser spots and $D$ is the long-axis length or the short-axis length, as shown in Fig. 2.22. $\eta_L$ and $\eta_S$ is the overlapping rates in the long-axis (X) direction and the short-axis (Y) direction respectively. The output pulse energy 40 J and elliptical spot (long-axis length 12 mm and the short-axis length 3 mm) were selected with overlapping rates $\eta_L = 30$, 50 and 70 % in the simulation and experiment. In the present work, the residual stress S11 along long-axis (X) direction was considered.

### 2.7.3 The Simulation and Verification of Surface Residual Stress Distribution

The benchmark simulation is to verify that the single and multiple impacts of LSP simulation produced similar residual stress magnitudes and distributions as experimentally observed by Yakimets et al. [79]. The benchmark simulation consisted of a single LSP located at the center of the workpiece as shown in Fig. 2.23, and the simulation condition is also shown in Fig. 2.23. The benchmark

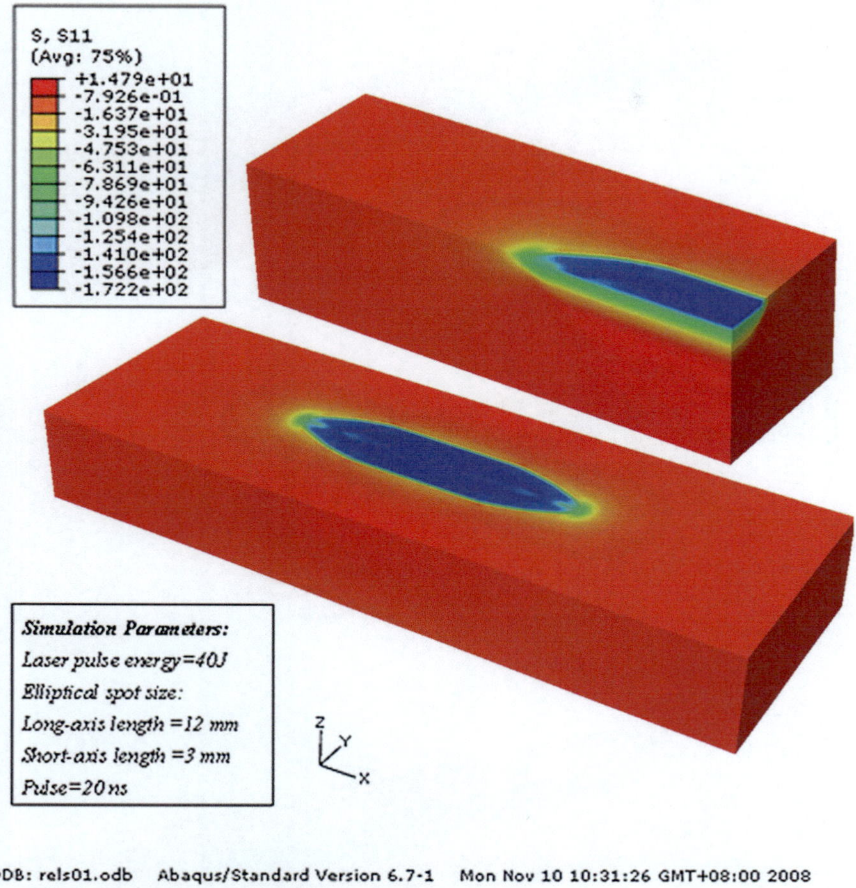


**Fig. 2.23** Benchmark simulation model of single LSP

simulation mesh consisted of C3D8R-type elements. Elements size is biased with a higher density of elements near the surface and gradually becoming less dense with increasing depth below the surface. The initial values of residual stress S11 were approximate in the zero-stress state.

Simulations were successively conducted with another two laser shocks on one side of the first treated zone by 30, 50 and 70 % overlapping rate along X direction. Figure 2.24 showed the distribution of surface residual stress S11 for different overlapping rates in the simulation with initial stress removed. The simulated residual stress distributions in the surface layer were given in Figs. 2.25a, b, and Fig. 2.25b was the enlarged partial view in overlapped regions. As shown in these figures, following observations can be made when compared with a single LSP impact.

The maximum magnitudes of compressive residual stress on the top surface were about -172 MPa for the first impact. However, it was greatly increased, after

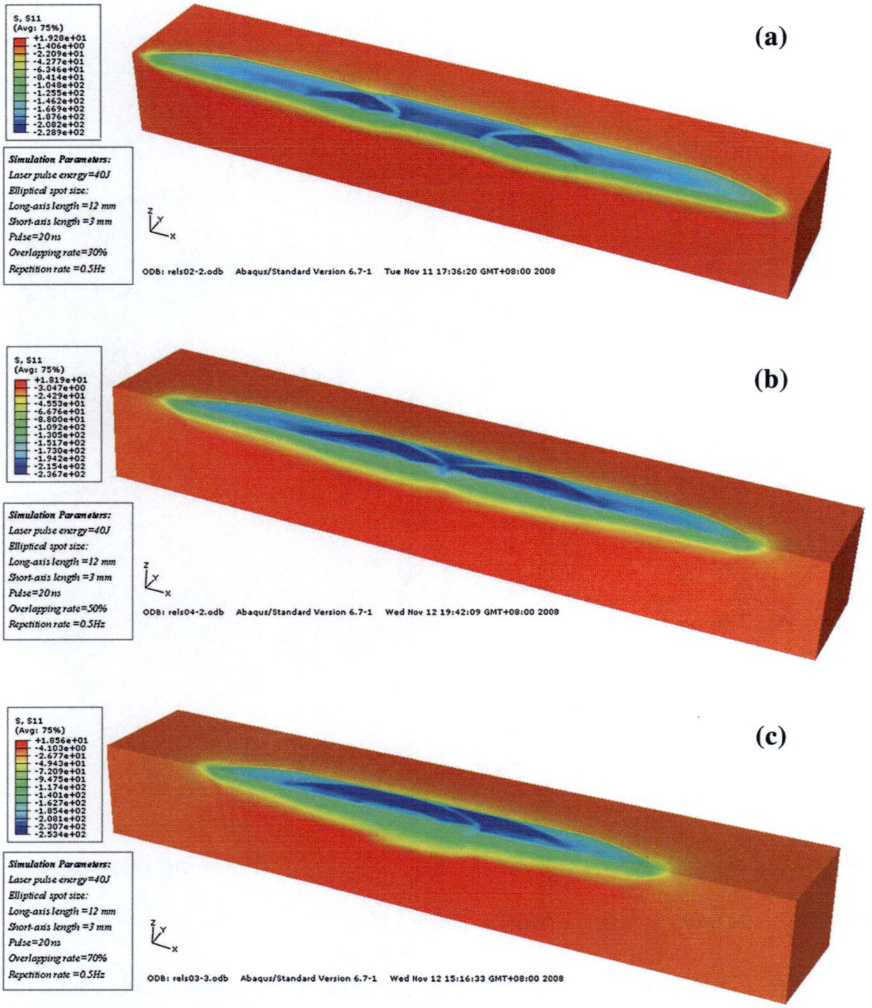


**Fig. 2.24** Simulation model of residual stresses S11 by LSP with over lapping rates $\eta_L$ **a** 30 %, **b** 50 %, **c** 70 %

two successive impacts on both sides. Obviously, it was about an increase of 33.7 % to −230 MPa for 30 and 50 % overlapping rate, and 47.1 % to −253 MPa for 70 % overlapping rate.

Under the 50 % overlapping rate, the fluctuation was about 4 %, and the good uniformity of surface residual stress along the symmetric plane was achieved. However, the fluctuation was increased by 23.5 % from −170 to −210 MPa for 30 % overlapping rate and by 15.4 % from −195 to −225 MPa for 70 % overlapping rate.

The results indicated that the fluctuation of LSP with 70 % overlapping rate was higher than that of LSP with 50 % overlapping rate due to the different

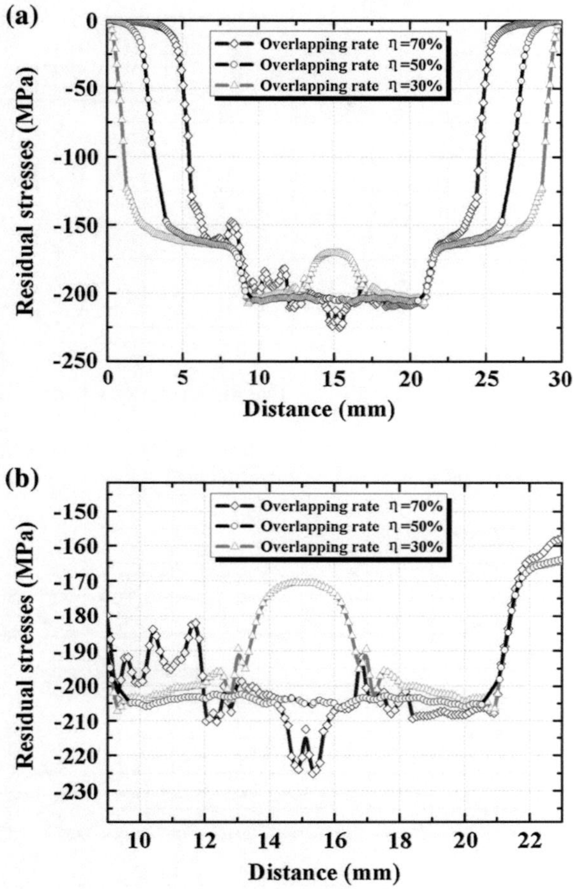


**Fig. 2.25** FEM simulation of surface residual stress S11 by LSP with different overlapping rates. **a** Along X direction on *top* surface, **b** the enlarged partial view in over lapped regions

shocked number at various regions. The simulated data was nonsymmetrical along the symmetric plane due to different shocked order.

Figure 2.26 showed the distribution of surface residual stress measured on the top surface along the symmetric plane with different overlapping rates. The measured results were in good agreement with the simulated data. Figure 2.27 showed the schematic diagram of overlapping elliptical spots along the center line with different overlapping rates $\eta_L$ (50, 70, and 80 %). From Fig. 2.27a, we can see that the regions along the symmetric plane were shocked two times while the overlapping rates $\eta_L$ was 50 %, and the distribution of surface residual stresses at the regions was uniform. While the overlapping rates $\eta_L$ is 70 %, it can be seen from Fig. 2.27b that the regions along the symmetric plane were shocked three times or four times, and the fluctuation of residual stress in the interesting region increased in comparison with that with the overlapping rate $\eta_L$ 50 %. If the overlapping rates $\eta_L$ is 80 %, it can be seen from Fig. 2.27c that the regions along the center line were shocked five times, and the homogeneity of surface residual

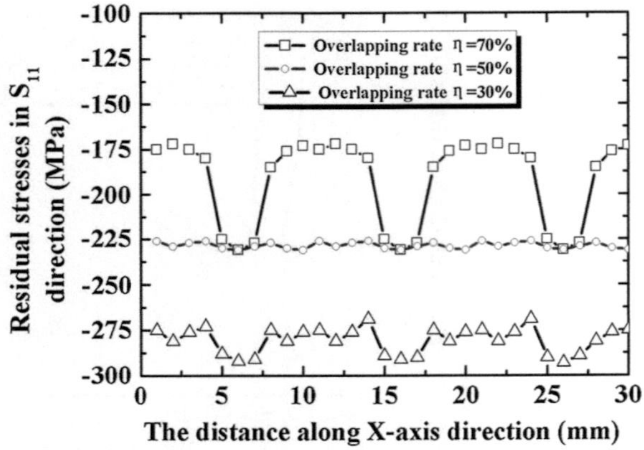


**Fig. 2.26** The distribution of surface residual stress measured on the top surface along the symmetric plane with different over lapping rates

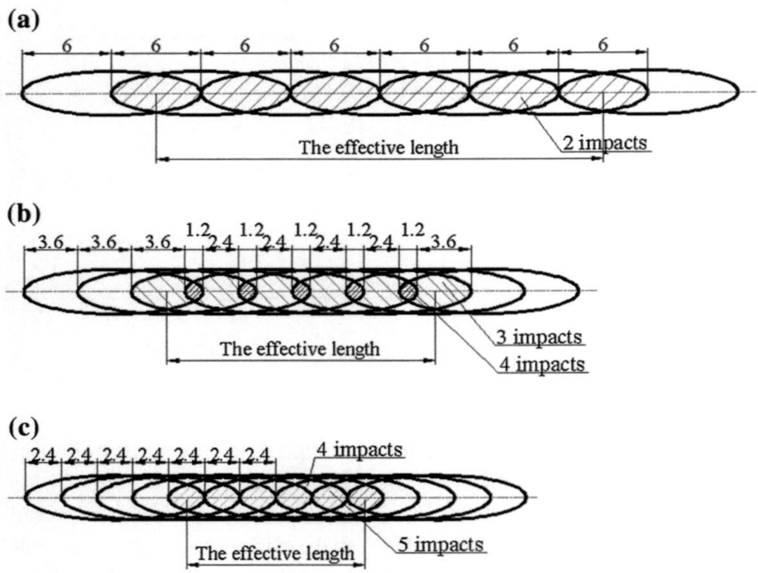


**Fig. 2.27** The schematic diagram of over lapping elliptical spots along the symmetric plane with different over lapping rates $\eta_L$. **a** 50 %, **b** 70 %, **c** 80 %

stresses at the regions was good, but the magnitude of surface residual stresses remain unchanged in comparison with that with the overlapping rate $\eta_L$ 50 %.

From the above studied results, we can make the following inferences. First, it is not the fact that the higher the overlapping rate is, the better the uniformity of the surface residual stress on the sample surface is. By compared to that with 50 % overlapping rate, the fluctuation of surface residual stress with 70 % overlapping rate is high. The higher the overlapping rate is and the more the shocked number is, the higher the magnitude of surface residual stress is. However, when the shocked number exceeds four times, the surface residual stress is kept to be the constant because the plastic deformation near the surface is almost saturated [80]. Hence, the increase of overlapping rate can enhance the values of surface residual stresses in repetitive-impacted regions within a certain impact-number range of 1–4 times, but can't effectively make it more uniform. Second, the higher the overlapping rate is, the smaller the effective shocked-length is. From Fig. 2.27, we can see that the effective lengths with the overlapping rate of 50, 70 and 80 % are 30, 18 and 12 mm under the same shocked-number, respectively. The bigger effective shocked-length can be obtained by using the smaller overlapping rate during LSP, thus the efficiency of the LSP can be improved.

### *2.7.4 The Simulation and Verification of Residual Stress Distribution in Depth Direction*

The simulated residual stresses were obtained from the surface element located at the center of the laser spot. A comparison of the simulated and measured residual stress values were shown in Fig. 2.28. Both the simulated and measured residual stresses were compressive, which are agreed with the trend.

Figure 2.29 showed the comparison of simulated residual stress S11 in depth direction with measured data. The simulated and measured in-depth residual stresses in the regions after 1 impact, 2 impacts and 3 impacts as functions of the distance from the surface were shown in Figs. 2.29a, b, respectively. It can be seen that the compressive residual stresses existed in the subsurface for both cases and the maximum magnitudes were located at the surface.

It is well known that the compressive residual stresses near the sample surface are generated due to the local plastic deformation after LSP. Both the simulated and measured residual stresses are on the whole consistent. When the distance from the surface is higher than the certain value (e.g., in the experiment it is about

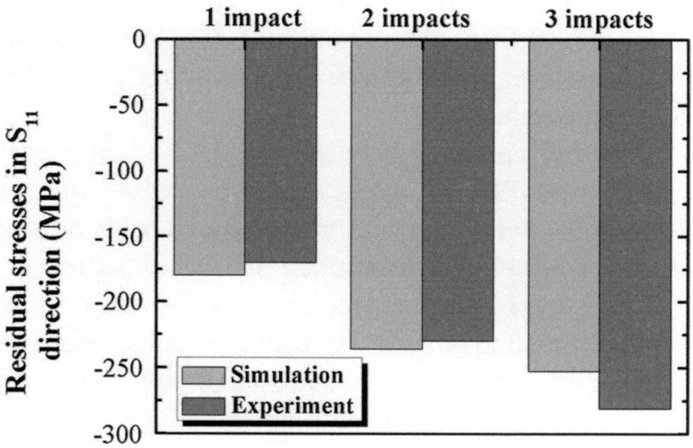


**Fig. 2.28** Comparison of simulated surface residual stress S11 with measured data

Fig. 2.29 Comparison of simulated in-depth residual stress S11 with measured data **a** simulated data, **b** measured data

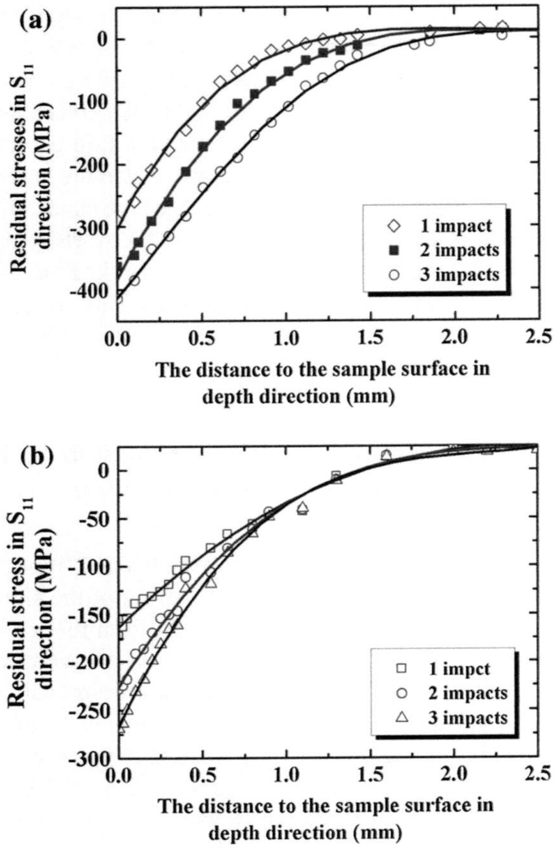


1.5 mm and in the simulation is about 2.0 mm in the present work), the residual stresses in the subsurface are almost not sensitive to the number of the laser impacts. However, when the distance from the surface is lower than the certain value, the number of the laser impacts has an important influence on the change of residual stress in depth direction. Hence, in order to increase the residual stress near the surface, the number of the laser impacts should be increased, but increasing the shock number cannot enough effectively to increase the plastically affected depth.

Some discrepancy between simulation and experiment may be due to several factors that different experimental procedure from the simulation. (1) The exact location of residual stress measurement with regard to the laser shocked region can't be accurately controlled in the experiment. (2) The magnitudes of the residual stress in depth direction were measured by using electropolishing material removal method, where the measured values need to be revised.

To obtain a homogeneous residual stress field, the overlapping rate should be optimized for the treatment. Laser impacts with 50 % overlapping rate can be used to obtain the surface residual stress field of good uniformity for LSP of LY2 Al alloy. In addition, in view of work efficiency, overlapping rate which is set to 50 % can be used to achieve longer effective shocked length. Hence, proper selection of overlapping rate is of great importance for the treatment of large area LSP, which can induce a better residual stress field in the sample to enhance the surface performance of the metal materials.

## 2.8 Summary

This chapter experimentally investigates and analyzes the effects of LSP on nano-hardness, elastic modulus and surface topography of the sample manufactured by LY2 Al alloy, and simulates and experimentally investigates the effects of overlapping rate on residual stress of LY2 Al alloy subjected to LSP with the elliptical spot. Some important conclusions can be summarized by the following statements:

(1) The values of nano-hardness and elastic modulus in the laser-shocked region and the laser-affected region were clearly larger than those in the non-shocked region. The contact depths in the laser-shocked region and the laser-affected region were lower than those in the non-shocked region, and the magnitude of nano-hardness in the laser-shocked region were greater from another point of view. The improvement of the elastic modulus and nano-hardness was favorable for improving FOD resistance of the components manufactured by LY2 Al alloy.
(2) The enhancement of LSP on nano-hardness and elastic modulus of LY2 Al alloy was mainly due to the grain refinement, the generation of high density dislocations and a small amount of twins.
(3) The initial surface topography plays a highly important role for LY2 Al alloy when treated by LSP. It will influence the surface topography, residual stress and micro-hardness.
(4) The surface roughness of LY2 Al alloy will tend to stable after one LSP impact. There is an ultimate value of about 0.58 μm for the surface roughness of LY2 Al alloy after multiple LSP impacts. The corresponding surface height distribution scatters from −2.5 to 2.5 μm.
(5) The increasing surface roughness will decrease the compressive residual stress when subjected to one LSP impact. When subjected to three LSP impacts, the surface residual stress of all the samples tends to be saturated and the saturated compressive residual stress is independent of surface roughness.
(6) LSP can improve micro-hardness of LY2 Al alloy. The increasing surface roughness will increase the micro-hardness after one LSP impact. However, the initial surface topography has no influence on the micro-hardness of LY2 after multiple LSP impacts.
(7) The single and multiple LSP FEM simulations have been developed and the simulated residual stresses in surface and depth direction agree well with the measured data in nature and trend.

(8) The largest stress magnitudes are located on the top surface of the LY2 Al sample, and the increase of the laser shocked-number can increase the magnitude of residual stress near the surface within a certain impact-number range of 1–4 times, but it is not effective to increase the affected depth.
(9) The overlapping rate should be optimized and selected carefully for large area LSP. An appropriate overlapping rate is beneficial to obtain a homogeneous residual stress field, and 50 % overlapping rate can be selected for the LSP treatment of the LY2 Al alloy due to the shocked effect and the work efficiency.

## References

1. Fairand, B. P., Wilcox, B. A., Gallagher, W. J., & Williams, D. N. (1972). Laser shock induced microstructural and mechanical property changes in 7075 Al. *Journal of Applied Physics, 43*, 3893–3895.
2. Peyre, P., Fabbro, R., Merrien, P., & Lieurade, H. P. (1996). Laser shock processing of Al alloys. Application to high cycle fatigue behaviour. *Materials Science and Engineering A, 210*, 102–113.
3. Ballard, P., Fournier, J., Fabbro, R., & Frelat, J. (1991). Residual stresses induced by laser-shocks. *Journal de Physique IV, C3*, 487–494.
4. Devaux, D., Fabbro, R., & Virmont, J. (1991). Generation of shock waves by laser–matter interaction in confined geometries. *Journal De Physique IV, 1*(C7), 179–182.
5. Peyre, P., & Fabbro, R. (1995). Electromagnetic gauge study of laser-induced shock waves in aluminium alloys. *Journal De Physique III France, 5*, 1953–1964.
6. Peyre, P., & Fabbro, R. (1995). Laser shock processing: a review of the physics and applications. *Optical and Quantum Electronics, 27*, 1213–1219.
7. Peyre, P., Merrien, P., Lieurade, H. P., & Fabbro, R. (1995). Laser induced shock waves as surface treatment for 7075–T7351 aluminium alloy. *Surface Engineering, 11*, 47–52.
8. Berthe, L., Fabbro, R., Peyre, P., Tollier, L., & Bartnicki, E. (1997). Shock waves from a water-confined laser-generated plasma. *Journal of Applied Physics, 82*, 2826–2832.
9. Vaccari, J. A. (1992). Laser shocking extends fatigue life. In J. A. Vaccari (Ed.), *Laser Technology* (pp. 62–64) American Machinist.
10. Dane, C. B., Hackel. L. A., Daly. J., & Harrison, J. (1997). High laser power for peening of metals enabling production technology. Advanced aerospace materials and processes Conference' 98. Tysons Corner,Virginia, June 15–18: 1998.
11. Tenaglia, R. D., & Lahrman, D. F. (2009). Shock tactics. *Nature Photonics, 3*, 267–269.
12. Fairand, B. P., Clauer, A. H., Jung, R. G., & Wilcox, B. A. (1974). Quantitative assessment of laser-induced stress waves generated at confined surfaces. *Applied Physics Letters, 25*, 431–433.
13. O'Keefe, J. D., & Skeen, C. H. (1972). Laser-induced stress-wave and impulse augmentation. *Applied Physics Letters, 21*, 464–466.
14. Hoffman, C. G. (1974). Laser–target interactions. *Journal of Applied Physics, 45*, 2125–2128.
15. Yang, L. C. (1974). Stress waves generated in thin metallic films by a Q-switched ruby laser. *Journal of Applied Physics, 45*, 2601–2607.
16. Ling, P., & Wight, C. A. (1995). Laser-generated shock waves in thin films of energetic materials. *Journal of Applied Physics, 78*, 7022–7025.
17. Couturier, S., Resseduier, M., Hallouin, M., Romain, J. P., & Bauer, F. (1996). Shock profile induced by short laser pulses. *Journal of Applied Physics, 79*, 9338–9342.

18. White, R. M. (1963). Elastic wave generation by electron bombardment or electromagnetic wave absorption. *Journal of Applied Physics, 34*, 2123–2124.
19. Skeen, C. H., & York, C. M. (1968). Laser-Induced blow-off phenomenon. *Applied Physics Letters, 12*, 369–371.
20. Clauer, A. H., Holbrook, J. H., & Fairand, B. P. (1981). Effects of laser induced shock waves on metals. In M. A. Meyers & L. E. Murr (Eds.), *Shock waves and high-strain-rate phenomena in metals* (pp. 675–702). New York: Plenum Publishing Corporation.
21. Fabbro, R., Fournier, J., Ballard, P., Devaux, D., & Virmont, J. (1990). Physical study of laser-produced plasma in confined geometry. *Journal of Applied Physics, 68*, 775–784.
22. Fairand, B. P., Wilcox, B. A., Gallagher, W. J., & Williams, D. N. (1972). Laser shock-induced microstructural and mechanical property changes in 7075 Al. *Journal of Applied Physics, 43*, 3893–3896.
23. Montross, C. S., Florea, V., & Bolger, J. A. (1999). Laser-induced shock wave generation and shock wave enhancement in basalt. *International Journal of Rock Mechanics and Mining Sciences, 36*, 849–855.
24. Fairand, B. P., Clauer, A. H., & Wilcox, B. A. (1977). Pulsed laser induced deformation in an Fe-3 Wt Pct Si alloy. *Metallurgical Transactions A, 8*, 119–125.
25. Clauer, A. H. (1996). Laser shock peening for fatigue resistance. In J. K. Gregory, H. J. Rack & D. Eylon (Eds.) *Surface performance of titanium* (pp. 217–230) Warrendale (PA): TMS.
26. Luong, H., & Hill, M. R. (2008). The effects of laser peening on high-cycle fatigue in 7085–T7651 Al alloy. *Materials Science and Engineering A, 477*, 208–216.
27. Zhang, Y. K., Zhang, X. R., Wang, X. D., Zhang, S. Y., Gao, C. Y., Zhou, J. Z., et al. (2001). Elastic properties modification in Al alloy induced by laser-shock processing. *Materials Science and Engineering A, 297*, 138–143.
28. Rubio-González, C., Ocaña, J. L., Gomez-Rosas, G., Molpeceres, C., Paredes, M., Banderas, A., et al. (2004). Effect of laser shock processing on fatigue crack growth and fracture toughness of 6061–T6 Al alloy. *Materials Science and Engineering A, 386*, 291–295.
29. Yang, J. M., Her, Y. C., Han, N., & Clauer, A. (2001). Laser shock peening on fatigue behavior of 2024–T3 Al alloy with fastener holes and stop holes. *Materials Science and Engineering A, 298*, 296–299.
30. Zhang, H., Zhang, Y. K., & Yu, C. Y. (1999). Surface treatment of Al alloy by laser shock processing. *Surface Engineering, 15*, 454–456.
31. Fourier, J. (1990). Mechanical effects induced by shock waves generated by high-energy laser pulses. *Materials Manufacturing Processes, 5*, 144–147.
32. Sánchez-Santana, U., Rubio-González, C., Gomez-Rosas, G., Ocana, J. L., Porro, J., & Morales, M. (2006). Wear and friction of 6061–T6 Al alloy treated by laser shock processing. *Wear, 260*, 847–854.
33. Zhang, Y. K., Lu, J. Z., Ren, X. D., Yao, H. B., & Yao, H. X. (2009). Effect of laser shock processing on the mechanical properties and fatigue lives of the turbojet engine blades manufactured by LY2 Al alloy. *Materials and Design, 30*(5), 1697–1703.
34. Clauer, A. H., Fairand, B. P., & Wilcox, B. A. (1977). Laser shock hardening of weld zones in Al alloys. *Metallurgical and Materials Transactions A: Physical, 8*, 1871–1876.
35. Lu, J. Z., Zhang, L., Feng, A. X., Jiang, Y. F., & Cheng, G. G. (2009). Effects of laser shock processing on mechanical properties of Fe–Ni alloy. *Materials and Design, 30*(9), 3673–3678.
36. Read, D. T., & Dally, J. W. (1993). A new method for measuring the strength and ductility of thin films. *Journal of Materials Research, 8*(7), 1542–1549.
37. Weihs, T. P., Hong, S., Bravman, J. C., & Nix, W. D. (1998). Mechanical deflection of cantilever micro beams: A new technique for testing the mechanical properties of thin films. *Journal of Materials Research, 3*(5), 931–942.
38. Doerner, M. F., Gardner, D. S., & Nix, W. D. (1986). A method for interpreting the data from depth-sensing indentation instruments. *Journal of Materials Research, 1*(4), 601–609.

39. Wang, X. W., Wang, J. Y., Wu, P., & Zhang, H. W. (2004). The investigation of internal friction and elastic modulus in surface nanostructured materials. *Materials Science and Engineering A, 370*, 158–162.
40. San, J. F., Wang, Z. C., Li, S. H., & Liu, J. J. (2006). Nano-hardness and wear properties of C-implanted Nylon 6. *Surface and Coatings Technology, 200*(18–19), 5245–5252.
41. Bhatt, R. T., Choi, S. R., Cosgriff, L. M., Fox, D. S., & Lee, K. N. (2008). Impact resistance of uncoated SiC/SiC composites. *Materials Science and Engineering A, 476*, 20–28.
42. Chen, X., Wang, R., Yao, N., Evans, A. G., Hutchinson, J. W., & Bruce, R. W. (2003). Foreign object damage in a thermal barrier system: mechanisms and simulations. *Materials Science and Engineering A, 352*, 221–231.
43. Montross, C. S., Ye, L., Wei, T., Clark, G., & Mai, Y. W. (2002). Laser shock processing and its effects on microstructure and properties of metal alloys: a review. *International Journal of Fatigue, 24*, 1021–1036.
44. Montross, C. S., Brandt, M., & Swain, M. V. (2001). Self-limiting hardness changes in laser peened 6061–t6 aluminium. *Surface Engineering, 17*, 477–482.
45. Montross, C. S., Florea, V., Brandt, M., & Swain, M. V. (2000). Subsurface properties of laser peened 6061–T6 Al weldments. *Surface Engineering, 16*, 116–121.
46. Devaux, D., Fabbro, R., & Tollier, L. (1993). Generation of shock waves by laser-induced plasma in confined geometry. *Journal of Applied Physics, 74*, 2268–2273.
47. Chu, J. P., Rigsbee, J. M., Banas′, G., & Elsayed-Ali, H. E. (1999). Laser-shock processing effects on surface microstructure and mechanical properties of low carbon steel. *Materials Science and Engineering A, 260*, 260–268.
48. Ganin, E., Komem, Y., & Rosen, A. (1978). Shock induced hardness in α-Iron. *Materials Science and Engineering A, 33*, 1–4.
49. Yilbas, B. S., & Arif, A. F. M. (2007). Laser shock processing of Al: model and experimental study. *Journal of Physics. D. Applied Physics, 40*, 6740–6747.
50. McEvily, A. J., Renauld, M., & Bao, H. (1997). Fatigue fracture-surface roughness and the K-opening level. *International Journal of Fatigue, 19*, 629–633.
51. Romeiro, F., Freitas, M., & Fonte, M. (2009). Fatigue crack growth with overloads/under loads: Interaction effects and surface roughness. *International Journal of Fatigue, 31*, 1889–1894.
52. Zhang, Y. K., Zhang, S. Y., & Zhang, X. R. (1997). Laser ultrasound velocity of material with a surface coating layer. *Surface and Coatings Technology, 92*, 104–109.
53. Rozmus-Górnikowska, M. (2010). Surface modifications of a Ti6Al4 V alloy by a laser shock processing. *Acta Physica Polonica A, 117*, 808–811.
54. Luong, H., & Hill, M. R. (2008). The effects of laser peening on high-cycle fatigue in 7085–T7651 Al alloy. *Materials Science and Engineering A, 477*, 208–216.
55. WYKO Vision Software Help Document.
56. Rozmus-Górnikowska, M. (2010). Surface modifications of a Ti6Al4 V alloy by a laser shock processing. *Acta Physica Polonica A, 5*, 117–120.
57. Berthe, L., Fabbro, R., Peyre, P., Tollier, L., & Bartnicki, E. (1997). Shock waves from a water-confined laser-generated plasma. *Journal of Applied Physics, 82*, 2826–2832.
58. Guo, Y. B., & Caslaru, R. (2011). Fabrication and characterization of micro dent arrays produced by laser shock peening on titanium Ti–6Al–4 V surfaces. *Journal of Materials Processing Technology, 211*, 729–736.
59. Clauer, A. H., Holbrook J. H., Fairand. B. P. (1981). In M. A. Meyers & L. E. Murr (Eds.). Sh*ock waves and high-strain-rate phenomena in metals* (pp. 675–702). New York: Plenum Publishing Corporation.
60. Yilbas, B. S., & Arif, A. F. (2007). Laser shock processing of Al: model and experimental study. *Journal of Physics D Applied Physics, 40*, 6740–6747.
61. Zhang, Y. K., Hu, C. L., Cai, L., Yang, J. C., & Zhang, X. R. (2001). Mechanism of improvement on fatigue life of metal by laser-excited shock waves. *Applied Physics A, 72*(2), 113–116.

62. Yang, C. H., Hodgson, P. D., & Liu, Q. C. (2008). Geometrical effects on residual stresses in 7050–T7451 Al alloy rods subject to laser shock peening. *Journal of Materials Processing Technology, 201*, 303–309.
63. Rubio-González, C., Ocaña, J. L., & Gomez-Rosas, G. (2004). Effect of laser shock processing on fatigue crack growth and fracture toughness of 6061–T6 Al alloy. *Materials Science and Engineering A, 386*, 291–295.
64. Zhang, Y. K., Gu, Y. Y., Zhang, X. Q., & Zhou, J. (2006). Study of mechanism of overlay acting on laser shock waves. *Journal of Applied Physics, 100*, 103517-1.
65. Marchi, C. S., Zaleski, T., Lee, S., Yang, N. Y., & Stuart, B. (2008). Effect of laser peening on the hydrogen compatibility of corrosion-resistant nickel alloy. *Scripta Materials, 58*(9), 782–785.
66. Zhang, Y. K., Zhang, S. Y., & Cai, L. (1997). Investigation of surface qualities of laser shock-processes zones and the effect on the fatigue life of Al alloy. *Surface and Coatings Technology, 92*, 104–107.
67. Warren, A. W., Guo, Y. B., & Chen, S. C. (2008). Massive parallel laser shock peening: Simulation, analysis, and validation. *International Journal of Fatigue, 30*(1), 188–197.
68. Dane, C. B., Hackel, L. A., Daly, J., & Harrisson, J. (1997). Laser peening of metals-enabling laser technology. *Advanced Materials Processes, 5*, 13–27.
69. Vaccari, J. A. (1992). Laser shocking extends fatigue life. *American Machine, 6*, 62–64.
70. Ding, K., & Ye, L. (2003). Three-dimensional dynamic finite element analysis of multiple laser shock peening processes. *Surface Engineering, 19*, 351–358.
71. Hu, Y. X., Yao, Z. Q., & Hu, J. (2006). 3-D FEM simulation of laser shock processing. *Surface and Coatings Technology, 201*, 1426–1435.
72. Nalla, R. K., Altenberger, I., Noster, U., Liu, G. Y., Scholtes, B., & Ritchie, R. O. (2003). On the influence of mechanical surface treatments-deep rolling and laser shock peening-on the fatigue behavior of Ti-6Al-4 V at ambient and elevated temperatures. *Materials Science and Engineering A, 355*, 216–230.
73. Nikitin, I., Sholtes, B., Maier, H. J., & Altenberger, I. (2004). High temperature fatigue behavior and residual stress stability of laser-shock peened and deep rolled. *Scripta Materials, 50*(10), 1345–1350.
74. Hu, Y. X., & Yao, Z. Q. (2008). FEM simulation of residual stresses induced by laser shock with overlapping laser spots. *Acta Metallurgica Sinica (English Letters), 21*(2), 125–132.
75. Evans, A. D., Bruno, G., King, A., Withers, P. J. (2002). Laser shock peening for aerospace Ti-6Al-4 V alloy: a residual stress study. Annual report of the institut Laue-Langevin (ILL), pp. 42–43.
76. Clauer, H. A., Walters, C. T., & Ford, S. C. (1983). The effects of laser shock processing on the fatigue properties of -T3 Al. In *Lasers in materials processing.* Metals Park (OH), American Society for Metals, pp 7–22.
77. ABAQUS, Inc. (2012). ABAQUS User's Manual, Ver. 6.12, Pawtucket, RI.
78. Ding, K., & Ye, L. (2003). FEM simulation of two sided laser shock peening of thin sections of Ti-6Al-4 V alloy. *Surface Engineering, 19*, 127–133.
79. Yakimets, I., Richard, C., Beranger, G., & Peyre, P. (2004). Laser peening processing effect on mechanical and tribological properties of rolling steel 100Cr6. *Wear, 256*(3–4), 311–320.
80. Evans A. D., Bruno G., King A., & Withers P. J. (2002) Laser shock peening for aerospace Ti–6Al–4 V alloy: a residual stress study. Annual Report of the Institut Laue-Langevin (ILL), Grenoble, France, pp. 42–43.

# Chapter 3
# Tensile Properties and Fatigue Lives of LY2 Al Alloy Subjected to Laser Shock Processing

**Abstract** This chapter presents the effects of strain rate on the tensile properties and fracture morphologies of LY2 Al alloy subjected to LSP. Special attentions are paid to the effects of LSP processing parameters on the fatigue life and fracture morphology. In addition, the influence mechanisms of different LSP parameters on Al alloy are systematically revealed.

## 3.1 Introduction

Since the deeper compressive residual stress value induced by LSP can significantly improve material properties and limit the growth and development of surface cracks [1–7], LSP is now emerging as a novel industrial treatment to improve the fatigue lives of alloys and steels [8, 9]. Hence, many of the proposed applications of LSP are aimed to improve the fatigue life and strength of the material through surface modification [10, 11]. There is a close relationship between the tensile properties and the fracture morphologies under different strain rates.

This chapter presents the effects of strain rate on the tensile properties and fracture morphologies of LY2 Al alloy subjected to LSP. Special attentions are paid to the effects of LSP processing parameters on the fatigue life and fracture morphology. In addition, the influence mechanisms of different LSP parameters on Al alloy are systematically revealed.

## 3.2 Tensile Properties Under Different Stain-Rates

Al alloy has been widely applied in the automobile and aircraft/aerospace industries due to its outstanding specific strength, electrical conductivity, thermal conductivity, easily forming and making, etc. However, the lower tensile resistance of Al alloy confines its wide application.

Y. Zhang et al., *Laser Shock Processing of FCC Metals*,
Springer Series in Materials Science 179, DOI: 10.1007/978-3-642-35674-2_3,

Considerable researches have been carried out to examine the effects of the strain-rate on the tensile properties of metal materials or mechanical property at a certain strain-rate by different treatment methods. Garci'a-Bernal et al. [12] have investigated the effect of five FSP (Friction Stir Processing) conditions on the high-temperature deformation properties of a continuous cast Al–Mg alloy at a strain-rate of $3 \times 10^{-2}\ s^{-1}$. Results showed that FSP can refine the grain size to less than 3 μm and enable the tensile elongation. Garcia-Infanta et al. [13] have found that 7,075 Al alloy has an upstanding elongation after high-pressure torsion up to five turns at 6 GPa. Boyce et al. [14] have examined the strain-rate sensitivity of four high-strength, high-toughness steels at strain-rates ranging from 0.0002 to 200 $s^{-1}$. Fan et al. [15] have investigated the effect of deformation behavior of an ultrafine-grained Al–Mg alloy at different strain-rates. Kaibyshev et al. [16] have investigated the high strain-rate superplasticity in an Al alloy subjected to simple thermomechanical processing. Han et al. [17] have investigated the strain-rate-dependent deformation behavior of bimodal 5,083 Al alloys processed by cryomilling and pointing out that crack blunting/bridging by ductile coarse grains plays a significant role at lower strain-rates. Most of the above researches reveal that the strain-rate has an important influence on the tensile properties and the fatigue lives of materials under different work conditions. In the aeronautical industry, key mechanical components are constructed from Al alloy. LSP can improve the fatigue life and strength of Al alloy, and reliable design of structural components requires an understanding of the dynamic mechanical properties and the relation between their tensile properties and facture morphologies at relevant strain-rates. However, in spite of numerous published studies on the influences of LSP on tensile properties at a certain strain-rate, little attention has been focused on the investigations into the dynamic tensile properties at relevant strain rates and the effects of strain rate on the ultimate tensile stress (UTS) of Al alloy subjected to LSP. Further studies of LSP on the tensile property for Al alloy are still necessary.

With the above background in mind, we focus on the tensile properties of the untreated and treated LY2 Al alloy samples by LSP at the broad strain-rate from 0.00001 to 0.1 $s^{-1}$. The engineering stress–strain curves are investigated, and the UTS (the flow stress) and strain rate sensitivity are also analyzed in both conditions. These topics discussed in the present work can provide some important insights on the engineering application of Al alloy subjected to LSP.

### 3.2.1 Experimental Material and Parameters

The tensile sample was cut into the dog-bone shaped tensile sample, whose dimension was shown in Fig. 3.1a. LY2 Al alloy was chosen for the metal tensile sample. To ensure the tensile data for comparison, fifty samples were cut from the same LY2 Al alloy plate.

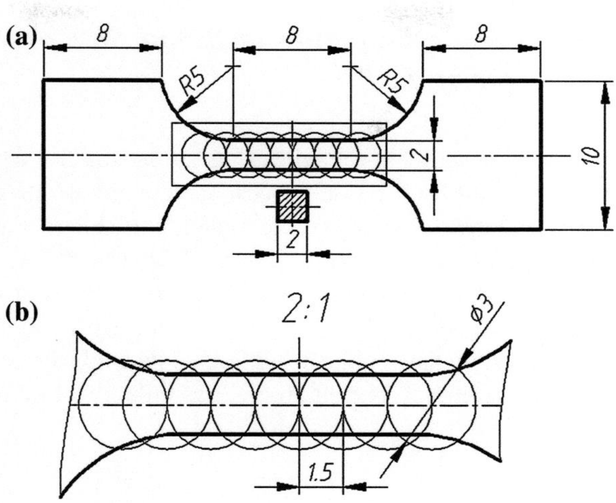


**Fig. 3.1** The dimensions of the tensile sample subjected to LSP (unit: mm). **a** The dimensions of the tensile sample, and **b** the partial enlargement drawing of the treated area subjected to LSP

All samples were cleaned in deionized water followed by ultrasonic vibration in ethanol to degrease their surfaces, of which twenty-five samples were treated shortly by LSP, and the other twenty-five samples were not been treated. The samples were treated by the laser shock wave along the center line, and the top surface of each sample was only treated, as schematically illustrated in Fig. 3.1a, while the partial enlarged drawing of the treated area subjected to LSP was shown in Fig. 3.1b. The diameter of the laser spot was 3 mm and the distance between the adjacent laser spots at the center line was 1.5 mm, namely, the overlapping rate was 50 % during LSP. In LSP, the shockwaves were induced by a Q-switched repetition-rate laser with a wavelength of 1,054 nm, a pulse duration of 20 ns and a pulse energy of around 3 J. Water with a thickness of 2 mm was used as the transparent confining layer and the 3 M Al foil with a thickness of 50 μm was used as the absorbing layer to protect the sample surface from the thermal effect.

The tensile properties of the treated and untreated samples were measured on a MTS880-10 servo-hydraulic material testing machine system at a broad strain-rate range of 0.00001–0.1 $s^{-1}$. For the treated and untreated samples, the tensile properties were measured five times for each test condition, thereby providing evidence of reproducibility. Under each test condition, the yield strength measured on the second test was varied by less than 2 % from the value measured on the first test. An extensometer with a gauge length of 25 mm was used to measure the strain during the test. The UTS and ductility (percent elongation) were evaluated.

Fractographic observations were performed using a 6,700F standard scanning field emission scanning electron microscope (FESEM) made in Japan. The microstructure of the near-surface region in the treated and untreated sample was characterized by using a JEM-2100 transmission electron microscope (TEM) operated at a voltage of 200 kV. All micrographs presented were bright-field images recorded under two-beam conditions.

### *3.2.2 Tensile Properties of Strain Rate on LY2 Al Alloy During LSP*

Figure 3.2a shows the engineering stress–strain curves of the untreated sample manufactured by LY2 Al alloy as a function of strain rate, and Fig. 3.2b is the partial enlarged drawing of the rectangle region in Fig. 3.2a. Figure 3.3a presents the engineering stress–strain curves of the treated LY2 Al alloy samples by LSP as a function of strain rate, and Fig. 3.3b is the partial enlarged drawing of the rectangle region in Fig. 3.3a. The engineering stress and the engineering strain approximately have a linear relation while engineering strain is smaller than 2.5 % in both conditions. The strain rate has little effect on the shape of the stress–strain

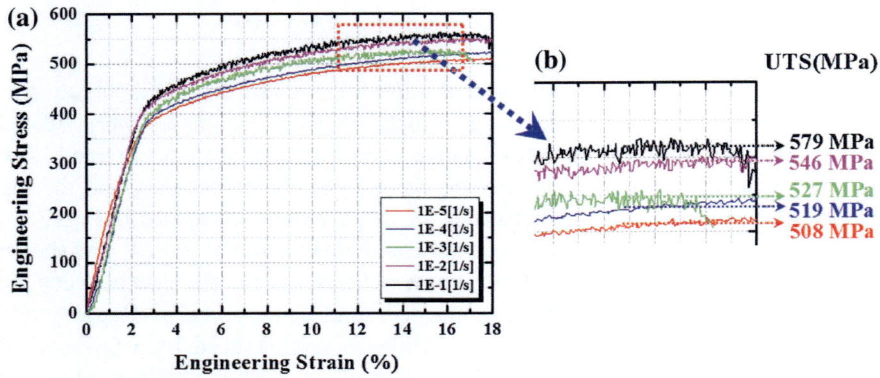


**Fig. 3.2** **a** The engineering stress–strain curves of the untreated sample manufactured by LY2 Al alloy as a function of strain rate, and **b** the high-magnification of the maximum value for the engineering stress at different strain rate

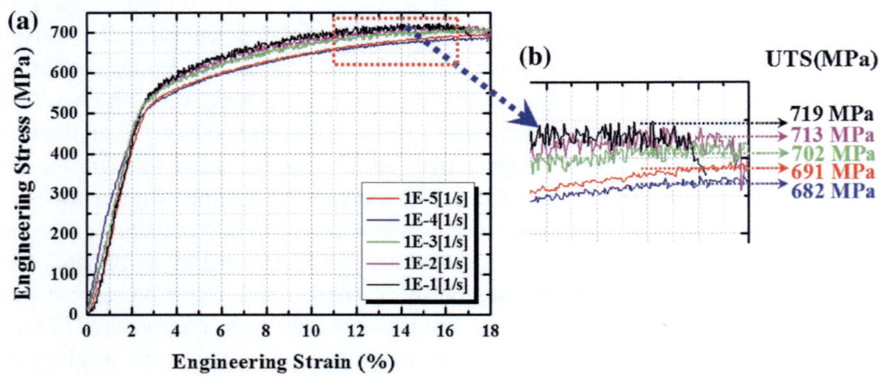


**Fig. 3.3** **a** The engineering stress–strain curves of the treated LY2 Al alloy samples by LSP as a function of strain rate, and **b** the high-magnification of the maximum value for the engineering stress at different strain rate

curve, and no significant change is observed in the slope of the linear region with the increase of the strain rate. The serrated flow (or the Portevin-Le Chatelier effect) can be clearly seen in the engineering stress–strain curves as a function of strain rate of the untreated and treated samples.

Figure 3.4a shows the relation between strain rate and the elongation of the untreated and treated samples, and Fig. 3.4b presents the relation between strain rate and the UTS of the untreated and treated samples, and Fig. 3.4c is the relation between strain rate and UTS of the treated sample. Both of the engineering stress–strain curves of the untreated and treated sample at the strain-rate ranging from 0.00001 to 0.1 $s^{-1}$ show a decrease in elongation (as shown in Fig. 3.4a). By comparing with the elongation of 16.7–19.3 % for the untreated sample, the

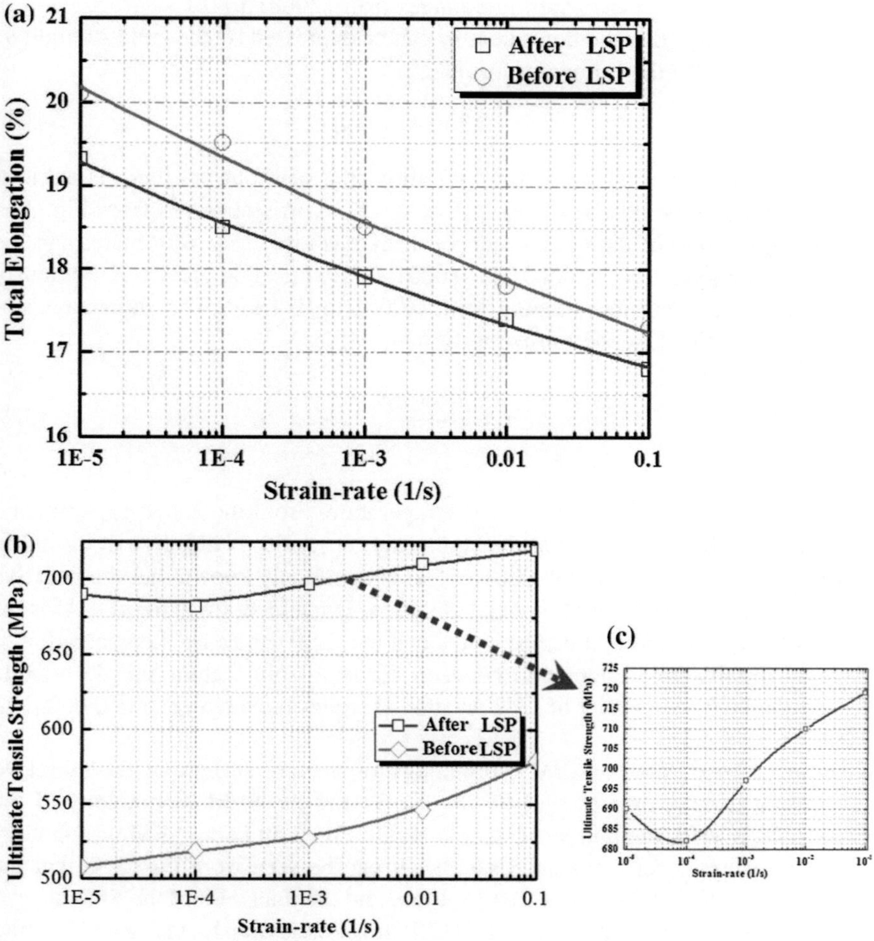


**Fig. 3.4** **a** The relations between strain rate and the elongation of the untreated and treated sample, **b** the relation between strain rate and the UTS of the untreated and treated sample, and **c** the relation between strain rate and UTS of the treated sample

treated sample presents the higher elongation of 17–20.2 %. Both elongations of the untreated and treated samples decrease gradually with the increase of the strain rate when the strain rate ranges from 0.00001 to 0.1 $s^{-1}$. Hence, it can be concluded that LSP can increase the elongation of the sample manufactured by LY2 Al alloy.

Figure 3.4b exhibits two UTS curves of the untreated and treated sample at the strain rate ranging from 0.00001 to 0.1 $s^{-1}$ at room temperature. When the strain rate increases from 0.00001 to 0.1 $s^{-1}$, the magnitude of UTS for the untreated sample increases gradually from 508 to 579 MPa. Whereas, it can be clearly seen from Fig. 3.4b to c that there is an obvious change in UTS when the strain rate ranges from 0.00001 to 0.1 $s^{-1}$, that is to say, there is a slight decrease in UTS when the strain rate ranges from 0.00001 to 0.0001 $s^{-1}$, while there is an obvious increase in UTS when the strain rate ranges from 0.0001 to 0.1 $s^{-1}$.

It is well known that the flow stress $\sigma_f$ can be expressed by the yield strength $\sigma_y$ and the UTS $\sigma_b$ as the following formula,

$$\sigma_f = 1.15 * (\sigma_y + \sigma_b)/2 \tag{3.1}$$

where the yield strength $\sigma_y$ of LY2 Al alloy is a constant, so there is a linear relation between the flow stress $\sigma_f$ and the UTS $\sigma_b$. It is observed from Fig. 3.4c that the flow stress of the treated sample decreases obviously with increasing the strain rate ranging from 0.00001 to 0.0001 $s^{-1}$, while that increases obviously with increasing strain rate ranging from 0.0001 to 0. 1 $s^{-1}$. The strain-rate sensitivity $m$ is defined by the following formula [15]

$$m = \left.\frac{\partial \log \sigma_f}{\partial \log \dot{\varepsilon}}\right|_{\varepsilon,T} \tag{3.2}$$

where $\dot{\varepsilon}$ is the strain-rate, and $T$ is the temperature. From the above experimental results, it can be calculated that the strain rate sensitivity is negative at the strain rate ranging from 0.00001 to 0.0001 $s^{-1}$. It is commonly known that the negative strain rate sensitivity is often associated with a dynamic strain aging (DSA) in Al alloy [18, 19]. Therefore, it can be concluded that LSP can cause the generation of DSA by the dynamic interaction between diffusing solute atom and the mobile dislocation in the SPD layer of LY2 Al alloy at strain-rate ranging from 0.00001 to 0.0001 $s^{-1}$.

Figure 3.5 exhibits the SEM morphology of fracture surface of the untreated and treated tensile samples by LSP. Figure 3.5a is the SEM morphology of the fracture surface for the untreated sample, while Fig. 3.5b is the SEM morphology of fracture surface for the treated sample. It can be seen from Fig. 3.5b that the depth of the SPD layer is about 210–220 μm, and the thickness of the SPD layer is uniform from the fracture surface after LSP impact. Obviously, the microstructure morphology of the SPD layer differs from that in the corresponding substrate. Figure 3.6 shows the TEM observations in the near-surface region of the untreated and treated sample. Figure 3.6a shows the typical TEM image of grains in the

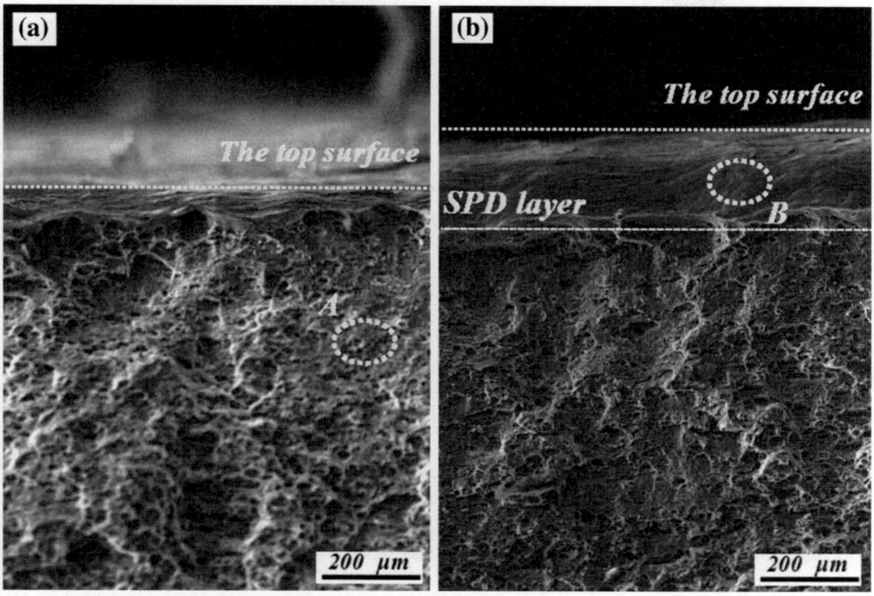


**Fig. 3.5** The SEM morphology of fracture surface of **a** the untreated, and **b** treated tensile samples by LSP

substrate at the depth of 200 μm from the top surface, it can be clearly seen from which that the size of the original grain is about 8–10 μm, as shown in the ellipses [A] in the Fig. 3.5a. Figure 3.6b shows the higher magnification of the SPD layer at the depth of 100 μm from the top surface, it can be clearly seen from which that the original coarse grains are subdivided into many sub-grains or refined grains with a width of 0.1–3 μm, as shown in the ellipses [B] in the Fig. 3.5b. Specially, it can be seen from the ellipse [C] in the Fig. 3.6b that grain size is about 100–200 nm.

The reaction between the laser shock wave and the Al alloy sample will be generated near the sample surface, leading to the generation of the dislocation activity and the microstructural deformation near the surface. After LSP, the grain is clearly refined, and the grain size is about 100–200 nm in the SPD layer. Many studied results have shown that the deformation of the FCC (Face Center Cubic) metal with grain size ranged from nanometer to sub-micron is mainly controlled by dislocation motion, and the increment of dislocation density greatly enhances the mechanical properties [20, 21]. The relation between the grain size and the macro-mechanical property of a ultrafine-grained AISI 304 SS indicates a strong improvement in both the tensile and fatigue resistance by grain refinement [22]. It is obtained from the above experimental results that the near-surface grains of the treated sample manufactured by LY2 Al alloy are refined by the generation of dislocation motion after LSP (as shown in Figs. 3.5 and 3.6), and the grain refinement has an obvious attribution to the improvement of the mechanical

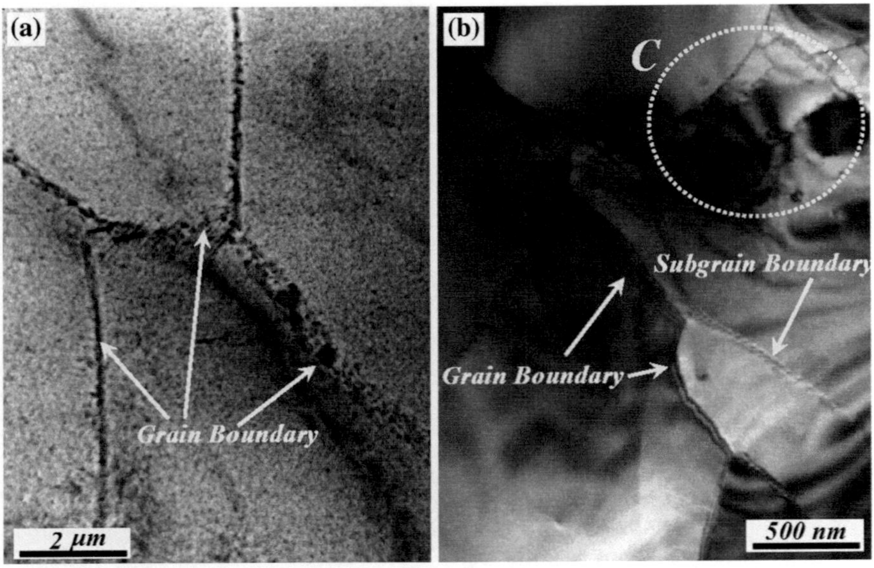


**Fig. 3.6** Typical TEM observations in the near-surface region of the untreated and treated sample. **a** The microstructure of the *ellipses A* in the Fig. 3.5a, and **b** the higher magnification of the *ellipses B* in the Fig. 3.5b

properties compared to the untreated sample [23]. In the previous papers, the refined Al-based alloy exhibited large tensile elongation, and grain refinement had an obvious attribution to the improvement of elongation in Al-based alloy [24–27]. Hence, the grain refinement in the shocked region after LSP has an obvious attribution to the improvement of the UTS and tensile elongation in LY2 Al alloy.

The strain rate plays an important role on the UTS (or flow stress) of metallic material. Generally speaking, the UTS of micrometer metal material increases significantly with increasing strain rate (when the strain rate is lower than $0.01\ s^{-1}$), while there is no significant change in UTS (when the strain rate is greater than $0.01\ s^{-1}$). However, the UTS value of the nano-structural and sub-micrometer metal material has an obvious change with increasing strain rate, that is to say, the UTS value of the sample decreases, reaches a local minimum and then increases within a broad strain rate. For instance, a ductile–brittle-ductile transition was founded in an electrodeposited 13 nm grain sized Ni–8.6 wt % Co alloy by changing the strain-rate. The strain rate range can be divided into two regions. At the low strain-rates of $1.04 \times 10^{-5}$ to $4.17 \times 10^{-2}\ s^{-1}$, the nano-crystalline alloy maintained the same level of UTS (about 2.0 GPa), while the plastic strain decreased dramatically from 4.2 to 1 % with increasing the strain-rate. As the strain rates further increased from $4.17 \times 10^{-2}$ to $1.04\ s^{-1}$, the UTS was obviously improved. The studied results gave the solid experimental evidence that the strain-rate would influence the UTS mechanism of nano-crystalline metals [28]. Han et al. [29] investigated the UTS and flow stress of the cryomilled refined structured 5083 Al alloy at different strain rates. Results showed that the UTS of the cryomilled nanostructured 5083 Al alloy decreases with increasing strain rate. The similar results can be seen in the paper [30]. LSP can generate the sub-micrometer grain in the shocked region of the treated sample, which may be the reason for the changing UTS curve (or flow stress) of the treated LY2 Al alloy.

## 3.3 Fractural Morphologies Under Different Stain-Rates

In the aeronautical industry, key mechanical components are constructed from Al alloy. LSP can improve the fatigue life and strength of Al alloy, and reliable design of structural components requires an understanding of the dynamic mechanical properties and the relation between their tensile properties and facture morphologies at relevant strain-rates. However, the effects of different strain-rate on the tension properties of Al alloy subjected to LSP are rarely reported, so the investigations into the dynamic mechanical properties of Al alloy subjected to LSP at relevant strain-rates have become essential.

With the above background in mind, the present investigation is aimed to examine the tensile properties and the fracture morphology of LY2 Al alloy subjected to LSP at different strain-rates. The stress–strain curves with strain-rate from 0.00001 to 0.1 $s^{-1}$ is investigated, and the relations between fracture morphology and strain-rate sensitivity is also analyzed. Furthermore, the fracture surfaces with different strain rate are processed by image-processing technique, which is used to deeply understand the mechanical behavior of LY2 Al alloy after LSP. These topics discussed in this paper can provide some important insights on the engineering application of Al alloy subjected to LSP.

### *3.3.1 Experimental Material and Parameters*

The tensile samples were cut into the dog-bone shaped tensile samples, and LY2 Al alloy was chosen for the metal tensile samples and its composition was listed in Table 2.3. The sample dimensions were shown in Fig. 3.1a, while Fig.3.1b was the partial enlarged drawing of the treated area subjected to LSP. To ensure the tensile data for comparison, all samples were cut from the same LY2 Al alloy plate. Prior to the laser shock, the samples were cleaned in deionized water followed by ultrasonic vibration in ethanol to degrease its surface, and then the samples were treated shortly by LSP.

Twenty-five samples, namely five samples for each test condition, were treated by the laser shock wave along the center line, as schematically illustrated in Fig. 3.1. The diameter of the laser spot was 3 mm and the distance between the adjacent laser spots at the center line was 1.5 mm, namely, the overlapping rate was 50 % during LSP. In LSP, the shockwaves were induced by a Q-switched repetition-rate laser with a wavelength of 1,054 nm, a pulse duration of about 20 ns and a pulse energy of around 3 J (laser power density of 4.3 GW/cm$^2$). Water with a thickness of about 2 mm was used as the transparent confining layer and the 3 M Al foil with a thickness of 50 μm was used as the absorbing layer to protect the sample surface from the thermal effect.

The tensile property of the treated sample subjected to LSP was measured on a MTS880-10 servo-hydraulic material testing machine system at a broad strain rate range from 0.00001 to 0.1 s$^{-1}$. Fractographic observations were performed using standard scanning field emission scanning electron microscope (FESEM) 6,700F made in Japan. Also, the total elongation of the tensile sample was determined using optical metrology on fiduciaries within the gage section.

### *3.3.2 Effects of Strain Rate on the Fractural Morphologies of LY2 Al Alloy by Laser Shock Processing*

Understanding the failure mechanisms under a wide range of loading conditions in metal materials would lead to not only better design methods but could also ways of enhancing the desired properties of metal materials. The fracture mechanism and the influence factors can be achieved by analyzing fracture morphologies at different strain-rates. The fracture morphologies are inspected in the SEM to observe strain-rate dependent changes in fractural morphology. When laser power density is 4.3 GW/cm$^2$, all high magnification fractural morphologies of LY2 Al alloy subjected to LSP at different strain-rates are included to exhibit the fracture morphologies as shown in Fig. 3.7a–e, and the corresponding microvoid networks of LY2 Al alloy samples subjected to LSP at different strain-rates are shown in Fig. 3.7a'–e'. It is seen from Fig. 3.7a–e that the typical dimple morphology appears on the fracture surfaces of LY2 Al alloy subjected to LSP. Similar to the common coarse-grained metal materials, the typical dimple coalescence morphology includes nucleating, growing, uniting, run-through and fracture of the tiny crack.

It is clear from Fig. 3.7a that there are plenty of dimples on the fracture surface, and the dimples were small and shallow at the strain-rate of 0.00001 s$^{-1}$. Figure 3.7b shows the dimples were symmetrical, and the cavity size is about 0.5–1 μm at the strain-rate of 0.0001 s$^{-1}$. From the fracture feature, it can be concluded that the fracture surfaces are vertical to the tensile direction because the positive tensile strength of the surface vertical to the tensile direction is maximal. When the strain-rate increases from 0.001 to 0.1 s$^{-1}$, the dimples become gradually larger and deeper, and it also can be seen some tearing ridges and impure particles on the fracture surface, as shown in Fig. 3.7b–e. In additional, Fig. 3.7d and e shows the dimple coalescence morphology on the fracture surfaces when the strain-rate increases from 0.01 to 0.1 s$^{-1}$.

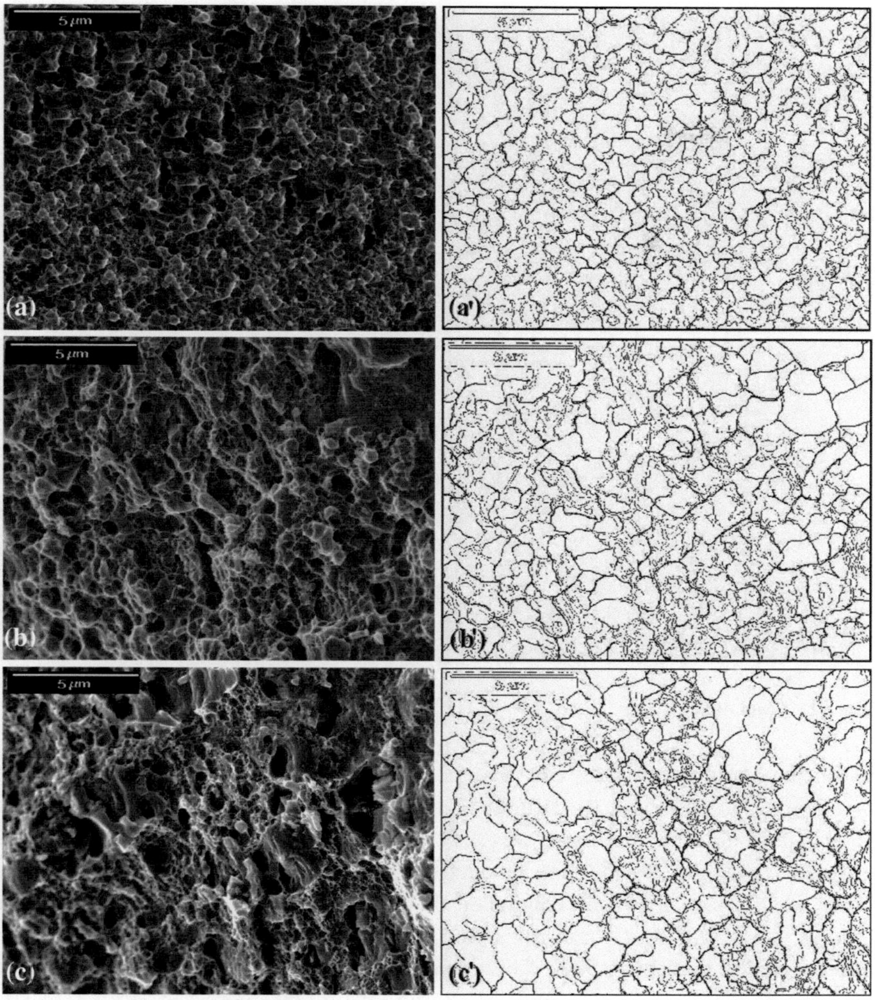


**Fig. 3.7** Fracture morphologies and the corresponding microvoid networks of LY2 Al alloy samples subjected to LSP at different strain-rates. **a–e** were the fracture morphologies, and **a'–e'** were the corresponding microvoid networks through image-processing. **a** and **a'** 1E − 5 1/s, **b** and **b'** 1E-4 1/s, **c** and **c'** 1E-3 1/s, **d** and **d'** 1E-2 1/s, **e** and **e'** 1E-1 1/s

The above fractographic examinations of the broken tensile samples reveals characteristic features like dimple coalescence indicative of ductile failure at the higher strain-rates, and the dimple coalescence takes place in parts containing inclusions or precipitates. It is well known that the presence of impure particles restricts dislocation mobility and has an obvious attribution to the initiation of crack nucleus and crack propagation. Generally speaking, the presence of dimple coalescence indicates clearly a transgranular ductile-mode fracture, and the dimple size in the fracture surface is a function of fatigue life, which has a direct relation

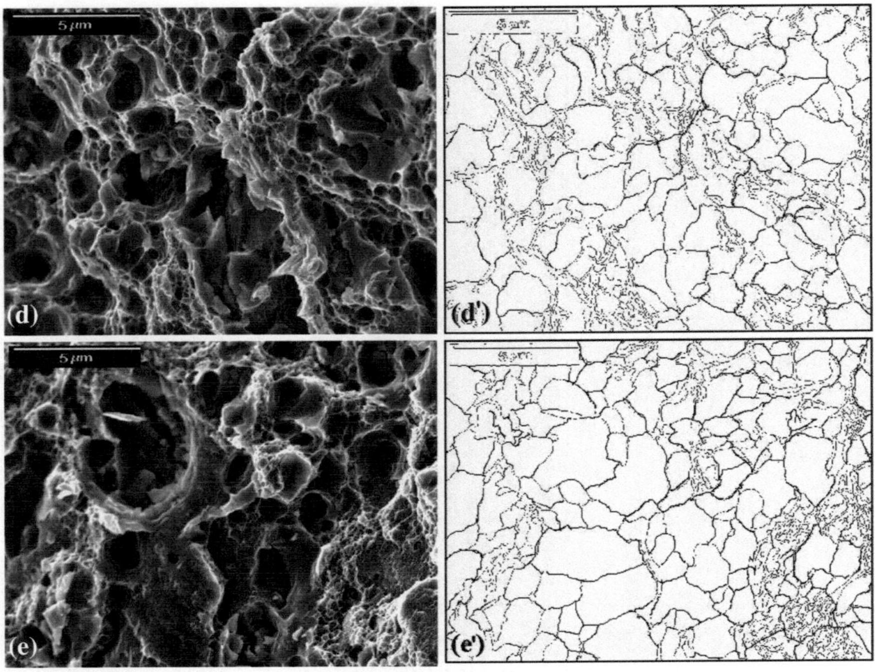


Fig. 3.7 (continued)

with the strain-rate [31]. The geometry of dimple and its correlation with mechanical properties in the structural materials have been investigated [31], and the results indicated that there was a systematic correlation between mechanical properties and the dimple size at ductile fracture with the strain-rate variation. From Fig. 3.7a–e, it can be concluded that there were obvious differences in the dimple size at different strain-rates, and the dimple size showed a notable strain-rate sensitivity after LSP. With the increment of the strain-rate, the dimple size and the dimple depth increase, and when the strain-rate increases from 0.001 to 0.1 $s^{-1}$, there was the dimple coalescence morphology on the fracture surfaces, and dimple becomes gradually larger and deeper. Hence, it is reasonable to estimate the mechanical properties of LY2 Al alloy with different strain-rates after LSP through fracture surface analysis of the dimple size.

In addition, it should be mentioned that if fracture morphology of a sample without LSP at different strain-rates are given by comparison, it will provide more evidence to explain that LSP give a strong impact on the strain rate sensitivity of LY2 Al alloy. More systematic experimental investigations are needed to demonstrate this effect.

## 3.4 Fatigue Life of LY2 Al Alloy by Laser Shock Processing Under Different Treatment Parameters

The turbojet engine blade manufactured by Ti alloy or Al alloy is an innovative aerospace component. However, the blade roots and edges experience fatigue loading during flight, and are subject to fretting fatigue where they contact the fan disc. In contemporary high bypass ratio engine designs, the blade may provide as much as 80 % of the total thrust. The LSP as a mechanical surface treatment has been reported to be used to repair the failed turbine components [32–34]. King et al. [34] investigated the effect of LSP on the fatigue strength and damage tolerance of the root of the blade manufactured by Ti-6Al-4 V alloy. Results showed that LSP can cause a local plastic deformation, resulting in a layer with the compressive residual stresses. The compressive stress could reach a depth of 2.0 mm. The existence of the compressive residual stress leads to the obviously increment of the fatigue lives of the samples. The similar results can be seen elsewhere [35–38]. Most of the above researches showed that the mechanical properties of blades had an important influence on the fatigue lives of them. However, to be effective, the compressive residual stresses that lead to the increment of the fatigue strength must be retained in service. For successful integration into turbine design, the process must be affordable and compatible with the manufacturing environment [38]. Moreover, most of the existing researches regarding the effect of LSP on the residual stress distributions and the fatigue performance of the blades are concentrated to the blades manufactured by Ti alloy. The similar researches on the Al alloy blade are relatively few. The aim of the present paper was to investigate the effect of LSP on the micro-hardness and residual stresses at the edges of the blade manufactured by LY2 Al alloy. The effects of the processing parameters, such as the number of shocks, used in LSP on the residual stresses and micro-hardness at the edge of the turbojet engine blade were investigated. The low cyclic fatigue performance on the samples cut from the edge of the blade was also evaluated. These topics discussed could provide some important insights on the maintenance and life-extension of the Al alloy blades.

### *3.4.1 Experimental Material and Parameters*

The edges of the turbojet engine blade manufactured by LY2 Al alloy was treated by using LSP, as shown in Fig. 3.8. In LSP, the shockwaves were induced by a Q-switched repetition-rate laser with a wavelength of 1,054 nm and a pulse of around 20 ns. The water with a thickness of 1–3 mm was used as the transparent confining layer and the 7,075 Al foil with a thickness of 0.1 mm was used as an absorbing layer to protect the blade surface from thermal effect. Laser energy was around 25 J and the power density at the blade surface was about 5 GW/cm$^2$.

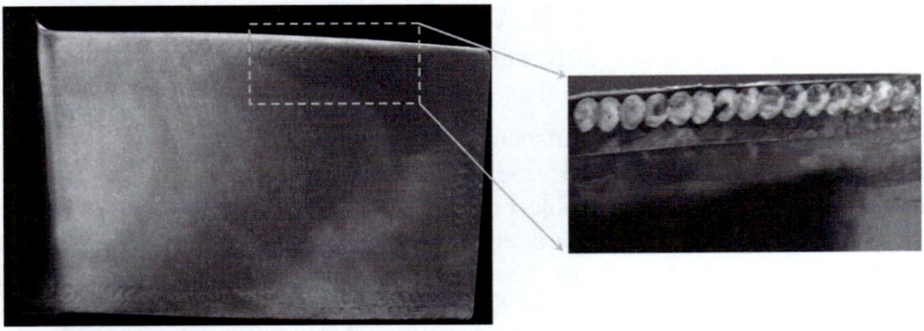

**Fig. 3.8** Photography of the turbojet engine blade by laser shock processing

The laser beam was continuously moved along the lines 1, and 2 parallel to the edge blade, as illustrated by the dash lines in Fig. 3.9. The distances from lines 1 and 2 to the edge of the blade were kept to be 2.5 and 5.5 mm, respectively, and the distance between the laser spots at line 1 and those at the line 2 was 3.0 mm. A denting mould was attached at another surface which was not subjected to the LSP to avoid the deformation of the blade during LSP.

The residual stresses at the surface of the sample and through the thickness direction after LSP were determined using X-ray diffraction. X-ray beam diameter was about 2 mm. X-ray source was Cr-K$\alpha$ ray and the diffraction plane was $\alpha$ phase (420) plane. In the stress calculation, the Poisson's ratio was set to be 0.25. The feed angle of the ladder scanning was 0.1° per sec. The scanning started angle and terminated angle were 159 and 168°, respectively. For the measurement of the residual stress along the depth direction, the electropolishing material removal method was used.

Micro-hardness tests were carried out by using a Vickers indenter under a 100 N load and 10 s hold time on as-polished and laser-processed regions. For each point, an average micro-hardness value was determined on the basis of the measured data from 10 indentations.

The samples with two notches were cut from the LY2 blade plate. Before fatigue testing, some samples were treated by LSP. The LSP procedure that had been used for the residual stress and hardness tests was the same as that of fatigue

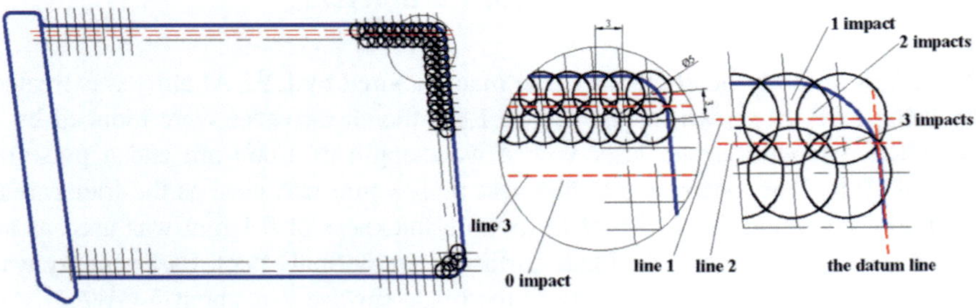


**Fig. 3.9** Schematic diagram of the shocking path

test. The regions subjected to LSP are schematically shown in Fig. 3.10. There were total four samples used in the fatigue tests to investigate the effect of the processing parameters used in LSP on the fatigue performance of the samples. For the first sample, only one side was treated by the single laser impact. The second sample was treated by the multiple laser impact at one side. However, for the third sample, two sides are treated by the single laser impact. The fatigue performance of the untreated sample was also evaluated for comparison. The axial low-cycle fatigue (LCF) tests were performed on a MTS880-10 servo-hydraulic material testing machine system at the room temperature. During LCF testing, the load ratio was maintained at $R = 0.2$ and the frequency of 1.5 Hz with a sine wave form was used. The maximum applied stress $\sigma_{max}$ was kept to be 255 MPa.

### 3.4.2 *Residual Stress Distribution Near the Edge of the Blade*

The residual stresses at the surface of the blade after LSP along the lines 1, 2, and 3, as functions of the distance from the datum line are shown in Fig. 3.11a. The location of the datum line is illustrated in Fig. 3.9. From Fig. 3.11a, it can be seen that the untreated regions are approximately in the zero-stress state, indicating that the effect of initial stress on the shock waves may be ignored. In the regions that are subjected to the multiple LSP with 3 laser impacts, as indicated in Fig. 3.9, the maximum residual stresses are around −250 MPa. However, in the regions after 2 laser impacts, the maximum residual stresses are −220 MPa. The compressive residual stress is only −150 MPa in the regions after 1 impact. It can be seen that when the number of the laser impacts increases from 1 to 2, the residual stress is

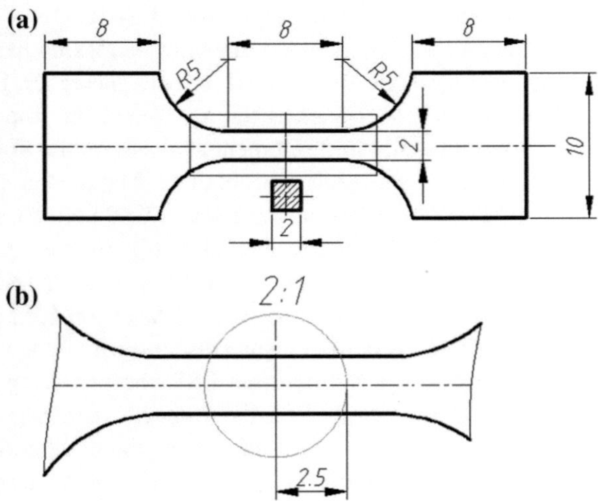


**Fig. 3.10** The dimensions of the tensile sample subjected to LSP (unit: mm). **a** The dimensions of the tensile sample, and **b** the partial enlargement drawing of the treated area subjected to LSP

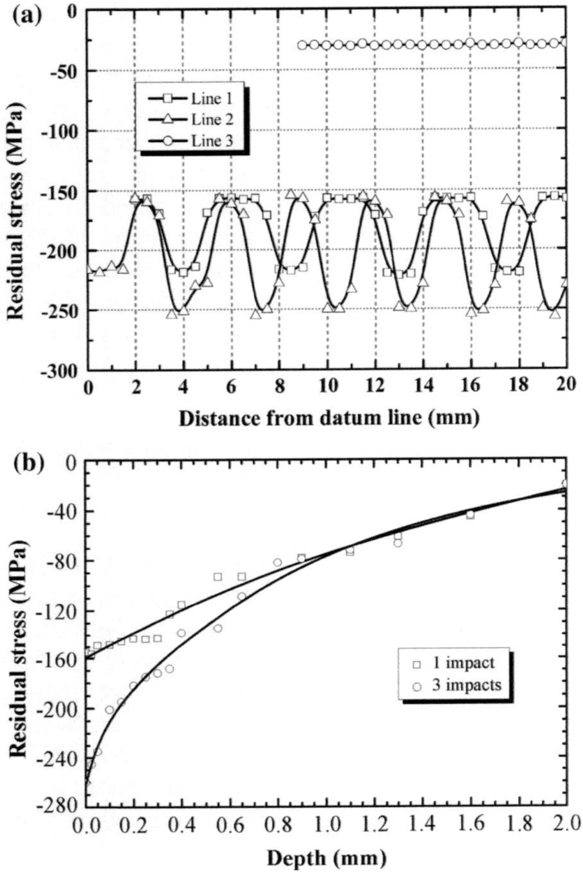


**Fig. 3.11** Residual stress distributions of the turbojet engine blade manufactured by LY2 Al alloy after LSP, **a** at the surface of the blade along the *lines 1*, *2*, and *3*, as functions of the distance from the *datum line*, and **b** in the regions after *1* impact and *3* impacts as functions of the distance from the surface are shown

increased by 46.67 %, whereas, when the number of the impact increases from 2 to 3, the residual stress is increased by 13.64 %.

Hence, it can be concluded that with increasing the number of the laser impacts in the multiple LSP, the residual stresses increases, but the increasing rate of the residual stresses decreases as the number of the impacts increases. Hu et al. [39] have performed 3-D finite element analyses on the residual stresses due to multiple LSP and the similar result was obtained. Masse et al. [40] investigated the effect of the number of the laser impacts used in LSP on the residual stress at the surface of the 4,340 steel sheet. Experimental results showed that for the untreated samples, the residual stress was about −500 MPa. After 1 laser impact, the residual stress reached −900 MPa. However, after 5 laser impacts, the residual stress was about −1,080 MPa. Hence, their results also indicated that with increasing the number of the laser impacts in the multiple LSP, the increasing rate of the residual stresses decreases as the number of the impacts increases. This phenomenon may be attributed to the fact that with increasing the number of the laser impacts, the depth of the plastic deformation is almost not changed, since the plastic deformation near the surface is almost saturated. The residual stresses in the regions after 1 impact and 3 impacts as functions of the distance from the surface are shown in Fig. 3.11b. It can be seen that, the compressive residual stresses exist in the subsurface for both cases and the maximum values are located at the surface. It is well known that the compressive residual stresses near the sample surface are generated due to the local plastic deformation after LSP. Hence, from Fig. 3.11b, it can also be concluded that the thickness of the plastic deformation layer near the surface is higher than 2.0 mm. When the distance from the surface is high than 1.1 mm, the residual stresses in the subsurface are almost not sensitive to the number of the laser impacts. However, when the distance from the surface is lower than 1.1 mm, the number of the laser impacts has an important influence on the residual stress magnitudes. Hence, in order to increase the residual stress near the surface, the number of the laser impacts should be increased.

### *3.4.3 Micro-Hardness Distribution Near the Edge of the Blade*

The micro-hardness at the surface of the blade after LSP along the lines 1, 2, and 3, as functions of the distance from the datum line are shown in Fig. 3.12a. By comparing the measured data of the micro-hardness long lines 1 and 2 and those along line 3, it can be seen that the micro-hardness of can be obviously increased after LSP. Moreover, with increasing the number of the laser impacts, the micro-hardness increases. This result agrees well with the experimental results obtained by Clauer et al. [41]. This phenomenon can be attributed to the following reason. After LSP, the reaction of the laser shock wave and the metal target will be generated near the target surface, leading to the generation of the dislocation and the microstructural deformation near the surface [42]. However, the increase in hardness was reported to be caused by an increase in the dislocation density with increasing laser impact repetitions [41].

The average micro-hardness at the line 1 is about 155 $HV_{0.2}$ and that at line 2 is 160 $HV_{0.2}$. At line 1, the maximum and minimum values of the micro-hardness are respectively located in the regions after two LSP impacts and one LSP impact. Similarly, at line 2, the maximum and minimum values of the micro-hardness are respectively located in the regions after three LSP impacts and two LSP impacts. The micro-hardness at the subsurface along the depth direction in the regions after one LSP impacts and three LSP impacts are shown in Fig. 3.12b. By comparing with the trends observed in Fig. 3.12b and those observed in Fig. 3.11b, some similar results can be obtained. When the distance from the surface is high than 2.0 mm, the micro-hardness in the subsurface are almost not sensitive to the number of the laser impacts. However, when the distance from the surface is lower than 2.0 mm, the number of the laser impacts has an important influence on the micro-hardness, i.e. with increasing the number of the laser impacts, the

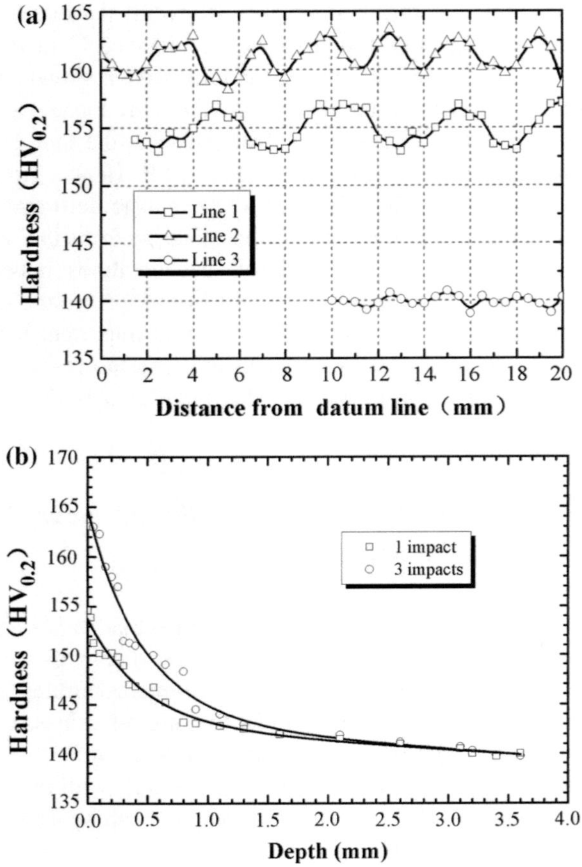


**Fig. 3.12** Micro-hardness distributions of the turbojet engine blade manufactured by LY2 Al alloy after LSP **a** at the surface of the blade after LSP along the *lines 1*, *2*, and *3*, as functions of the distance from the *datum line*, and **b** at the subsurface along the depth direction in the regions after one LSP impact and three LSP impacts

micro-hardness increases. Generally, a plastic deformation layer will be generated near the target surface. However, the depth of the plastic layer is influenced by the processing parameters used in LSP. Underneath the plastic layer, the microstructure of the metal target is almost not influenced by LSP. In such a case, in the region underneath the plastic layer, the micro-hardness will almost not changed due to LSP. But, in the plastic layer, the micro-hardness will be increased due to the increase of the dislocation density after LSP. Moreover, with increasing number of the laser impacts, the dislocation density increases [41].

### 3.4.4 *Effect of LSP Processing Parameters on the Fracture Morphology*

Figure 3.13 shows the SEM photos of fatigue fracture morphology of non-LSPed and LSPed LY2 Al alloy samples, respectively. It can be seen that the both fatigue fractures consist of three typical areas, i.e. fatigue source, fatigue crack growth area and final rupture. The initiation location of the fatigue source was closely related to the microscopic defects, surfaces quality and residual stress state of samples. From these two images, it can be seen that the significantly difference can be revealed by comparing the fatigue source of non-LSPed and LSPed samples. Figure 3.13a shows the fatigue fracture of non-LSPed sample, presenting typical features of the initiation of fatigue crack and consequently propagation. The site "A" in the left-upper corner of Fig. 3.13a is the fatigue source where the white arrow is propagation direction, and the area included in the white line is the fatigue cracks growth region. It can be seen from the fatigue source that fatigue cracks propagate in a radiation manner to the whole area, and assumed some cleavage and brittle fatigue fracture morphology, which is consistent with the fatigue behavior of high strength Al alloy.

As shown in Fig. 3.13b, there is a distance from the fatigue source to the sample's surface of the LSPed sample, and in this case the distance is nearly equivalent to the depth of the strengthened layer by LSP. Due to the great compressive residual stress of the sample surface induced by LSP, it significantly and effectively reduces stress concentration at the corner, and balances the peak value of tensile stress there. Meanwhile, the superimposition of compressive residual stress at sample surface and the applied load stress actually reduces the total stress of the sample, which caused the total surface stress loaded on the sample be well below the fatigue limit in the strengthened layers, so the fatigue crack source moved towards inner part of the strengthened layers. Although the fatigue cracks growth also assumes the scalloped radial manner, the propagation of the fatigue cracks towards the surface direction was hindered by the compressive residual stress at the surface, as shown in Fig. 3.13b.

Enforced under the cycled loads, fatigue crack growth develops into the second stage. At the beginning of this stage, there appeared the typical fatigue striations on fracture which are perpendicular to the direction of crack propagation. Because

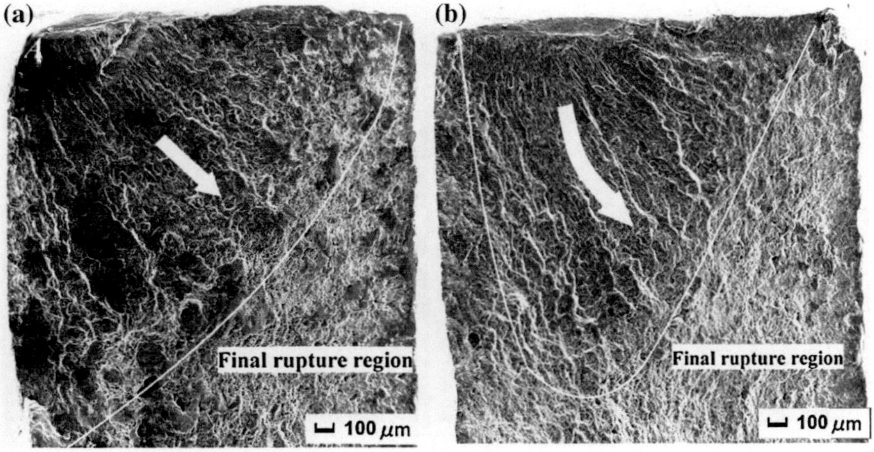


**Fig. 3.13** The morphology of the fatigue fracture of **a** non-LSPed sample and **b** LSPed sample

the existence of grain boundaries and the second phase particles in this material, sometimes the propagating direction has a few deviation at an angle from the main direction, as shown in Fig. 3.14.

Figure 3.14a shows the fatigue striations propagating situation of the non-LSPed sample, it can be seen that fatigue striations present a parallel manner and with the progressive spacing. Their crack growth rate da/dN reached $1.3 \times 10^{-3}$ mm cycle$^{-1}$, which is relative larger, even the second cracks can be clearly observed at this stage. Compared to Fig. 3.14a, the crack growth rate of the LSPed sample (as shown in Fig. 3.14b) is much slower than the non-LSPed sample, and the interval of the fatigue striations is narrower as well. Estimated according to the scale attached in Fig. 3.14, the width of the fatigue striations of the LSPed sample is only 1/4–1/5 of the non-LSPed sample for each progress of the fatigue cycle.

In addition, as shown in Fig. 3.14b, the edge of the fatigue striation on the LSPed sample presents bending and curve, just as cloud-like stripe pattern, and sometimes even difficult to distinguish between these striations. This is due to the following reasons. On the one hand, the fatigue striations are relatively flat and shadow at the initial period of the second stage of crack propagating. On the other hand, strong compressive residual stress plays an important role to minimize the open width of those striations during each progress of the fatigue cycle. Thus, in the fatigue striations growth process with each iteration, their opening amplitude is inevitably smaller than the non-LSPed sample, fatigue striation eventually become squeezed a cloud-like stripe pattern. Therefore, the fatigue striations cannot be clearly recognized, while those conventional samples left clear fatigue striations.

Figure 3.15 shows the typical images of fatigue crack propagation at the latter half period in the second stage. As the figures shown, fatigue crack growth rate of the non-LSPed sample significantly speeded up, the crack growth rate da/dN reached $3.0 \times 10^{-3}$ mm cycle$^{-1}$ above, moreover, the second cracks between striations became more remarkably, which indicated that the fatigue damage

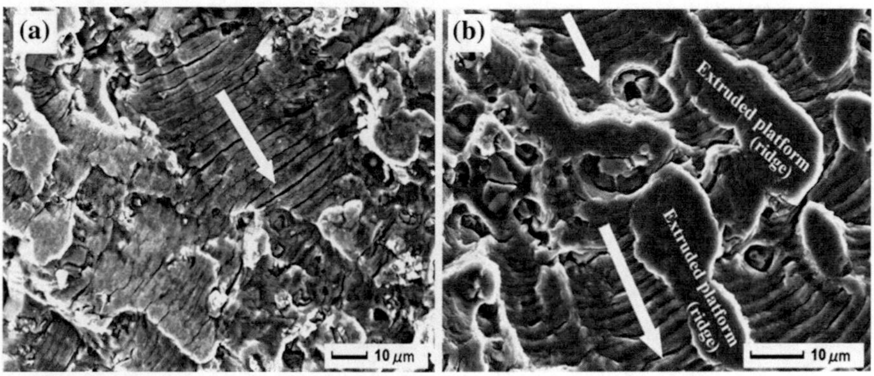


**Fig. 3.14** The SEM images at the beginning of stage II in fatigue crack propagation zone. **a** Fatigue striations of non-LSPed sample; **b** fatigue striations of LSPed sample

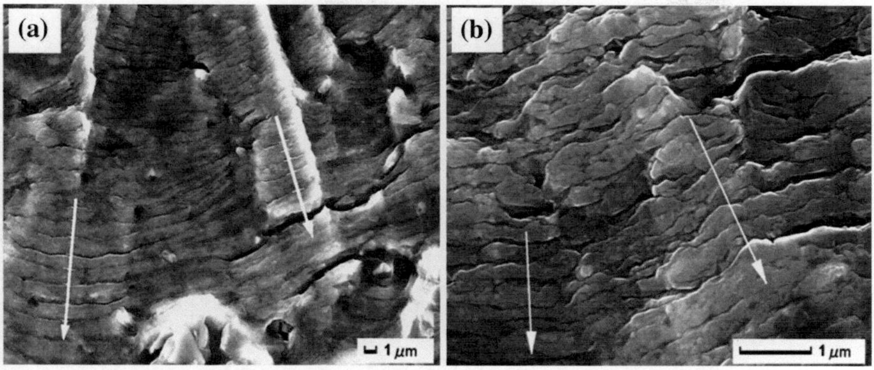


**Fig. 3.15** The SEM images at the ending of stage II in fatigue crack propagation zone. **a** Fatigue striations of non-LSPed sample, and **b** fatigue striations of LSPed sample

penetrated into the deeper inner part of the material. While the fatigue striation of the LSPed sample has also become wider, but propagated more stably, no second cracks observed. It should be noted that, on the fatigue fracture of the LSPed sample, some cloud-like platforms at same height were observed, as shown in Fig. 3.15b. This is because the peak-to-peak, valley-to-valley set vis-à-vis respectively on both sides of the fracture during the fatigue cracks grew forwards, therefore the cloud-like extruded ridges formed, which corresponded to the continual squeeze and grind each other between peaks. The fatigue striations, on these cloud-like extruded ridges, are not so easy to identify clearly, which becomes a typical kind of fatigue striation propagating morphology with non-crystallography feature, and should be an important characteristics of low cycle fatigue behavior for the Al alloy laser shock peened. It should also be noted that at this stage the fatigue striations at valleys are still very clear, and the crack growth rate da/dN remained less than $0.25 \times 10^{-3}$ mm cycle$^{-1}$, which exhibits the better fatigue life of the LSPed sample.

In this experiment of low cycle fatigue the applied loading stress is generally greater than the yield strength of test material, thus the test material has a certain amount of plastic deformation in each cycle, so the material exists in the state of elastic–plastic strain, the induced strain should include elasticity component and plastic component. When the ratio $\sigma_b/\sigma_{0.2}$ reaches 1.4 or so, the material will exhibit cyclic hardening effect [43]. It has been observed that, from the squeezed cloud-like extruded ridges, the test material underwent serious plastic strain in the fatigue proceeding, the deformation resistance of the Al alloy makes the material possess of the effect of cyclic hardening. This kind of enhancement effect would effectively inhibit the occurrence of the second cracks, and improve the fatigue performance of the material.

Figure 3.16 exhibits the SEM images of fracture morphology for the LSPed and non-LSPed samples. As shown in the figures, both the two fracture morphologies have the tensile fracture characteristics. Because the fatigue cracks propagation

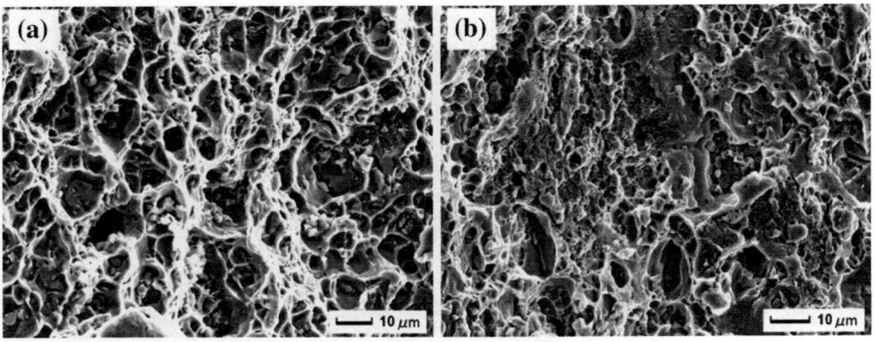


**Fig. 3.16** The SEM images of fracture morphology in the **a** non-LSPed and **b** LSPed samples

area is actually equal to a pre-installed crevice, the fracture surface renders the Griffith fracture characteristics. The final rupture region is mainly composed of dimples with different sizes, and towards an identical side of the fracture, presenting the fracture characteristics of tough materials. The most obvious difference of both two fractures is the real stress differential due to the difference of fatigue striation propagation area and the effects of residual stresses. The real final rupture stress of the non-LSPed sample should be larger than that of the LSPed sample. Therefore for the former, the final breaking speed is higher at the fracture surface rendering many uniform-dimples, as shown in Fig. 3.16a, whereas on the LSPed samples, owing to the role of residual stress, eventually the breaking speed would slow down, on the fracture surfaces showing some crystalline features of the raw materials, as shown in Fig. 3.16b, many tiny fracture dispersed small dimples observed on the figure, which is in an agreement with the microstructure of the material.

### *3.4.5 Effect of LSP Processing Parameters on the Fatigue Performance*

The fatigue lives of the treated samples using LSP with different parameters and the untreated sample are shown in Fig. 3.17. It can be seen that the sample treated by the two impacts in LSP at the one side exhibits the highest fatigue life. By comparing with the untreated sample, the fatigue lives of the samples are respectively increased by 131.4 and 132.5 % after the single impact and two impacts in LSP. The similar results can be seen in the paper [40]. This phenomenon may be attributed to the following reasons. First, the material below the irradiated surface is submitted to an elasto-plastic wave generating uniaxial plastic strain. The surrounding material is opposed to that strain and therefore induces, after the interaction, biaxial compressive residual stresses [44]. The compressive residual stresses with the values of several hundred MPa will be generated near the sample surface after LSP. It is well known that as fatigue cracks mostly originate

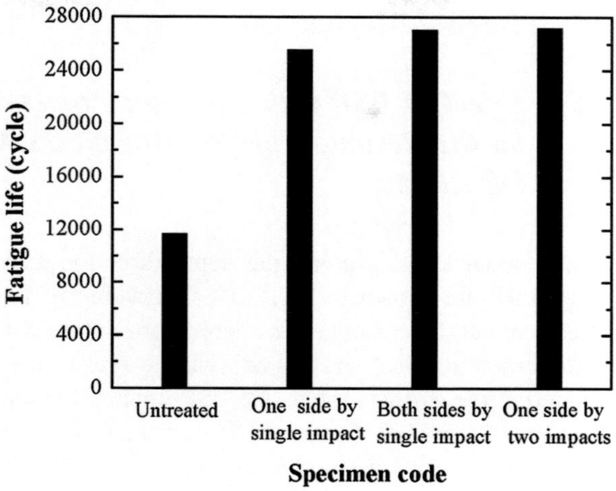


**Fig. 3.17** Fatigue properties comparison with different LSP parameters, the maximum applied stress $\sigma_{max}$ was kept to be 255 MPa

at the surface of materials, the fatigue behavior of mechanical parts depends strongly on their mechanical surface states, and a compressive surface layer prevents crack opening and growth and therefore has a beneficial effect on the fatigue performance [45]. Experimental result by Chien et al. [46] showed that the compressive stress at the surface would inhibit the crack initiation and growth during the LCF testing. Considering that fatigue crack nucleation is a surface phenomenon associated to the residual stress state on surface and subsurface, compressive residual stresses in the surface layers can increase fatigue performance [47, 48], and residual stress resulting from the plastic deformation of the surface layers of the material have a straight relation with the position of the crack source and, as a consequence, with the fatigue limit [49]. Hence, the compressive stresses at the surface may have an obvious attribution to the improvement of the fatigue performance of the sample after LSP.

Second, by comparing with the untreated sample, the surface roughness of the sample is relatively low. Since the high surface roughness may induce numerous stress concentrations, the low fatigue lives of the untreated sample may be due to the high surface roughness. However, in Fig. 3.18, it also can be found that the number of the laser impacts almost has no influence on the fatigue lives of the samples. Moreover, by comparing with the samples treated by LSP at one side, the fatigue life of the sample treated by LSP at two sides is relatively low. Hence, further work should be carried out to investigate the effect of the processing parameters in LSP on the fatigue lives of the samples.

## 3.5 Influence Mechanisms

### *3.5.1 Effect of LSP with Different Processing Parameters on the Residual Stress Distribution Along the Depth Direction*

The residual stresses along the depth direction after single-sided LSP and two-sided LSP are shown in Fig. 3.18. The value of the residual stress without any treatment can be considered as zero approximately. Firstly, it can be seen that the compressive residual stresses exist in the subsurface and the maximum values are located at the surface. After one LSP impact, the maximum compressive residual stress is −150 MPa. Subjected to two LSP impacts on one side, the maximum compressive residual stress is around −220 MPa. However, after two-sided LSP impacts, the maximum compressive residual stress is −195 MPa. The maximum value of compressive residual stress of the second group is increased by 46.67 % than that of the first group. While the maximum value of compressive residual stress of the third group is lower than that of the second group, but higher than the first group. Hence, it can be concluded that with increasing the LSP impact times, the maximum value of compressive residual stresses increases, but the effects obtained by two-sided LSP is not better than that by two LSP impacts on one side.

Secondly, it can be also found from Fig. 3.18 that the tensile residual stresses are generated at the mid-plane of the samples. After one LSP impact, the maximum value of tensile residual stress is 30 MPa at the depth of 1.0 mm. After two LSP impacts on one side, the maximum value of tensile residual stress is around 29.9 MPa at the depth of 1.0 mm from the top surface. However, after two-sided LSP impacts, the maximum value of tensile residual stress is 70 MPa at the depth of 0.70 mm from the top surface. Then the tensile residual stress is changed into the compressive stress of −1 MPa at the middle of the depth, 1.0 mm, from the top

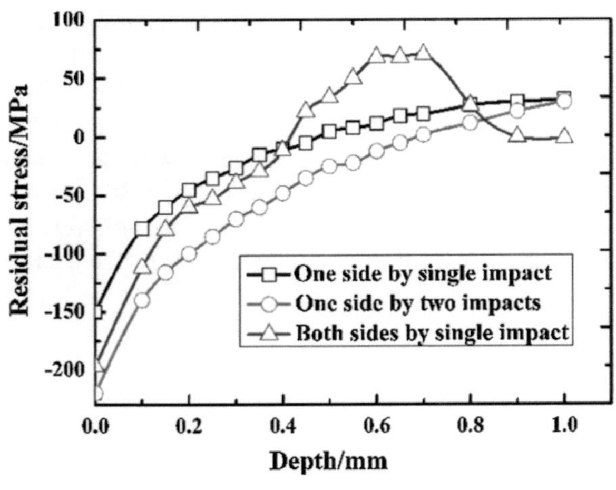


**Fig. 3.18** Residual stresses of different processing parameters along the depth direction

surface. But the compressive residual stress is much lower than the tensile residual stress. The values of the tensile residual stresses gained by single-sided LSP are almost the same, but lower than that by two-sided LSP. Between the surface and the mid-plane, the curve of the third group in Fig. 3.18 is a little steeper among the three groups. Due to the propagation of the shock waves from two sides of the samples, the two reverse shock waves encounter and overlap, the compressive stresses are cancelled by the tensile stresses. The thinner the samples are, the more obvious the phenomenon is. As for it, Ding et al. [50] have performed FEM simulation analyses on the residual stresses along three different thicknesses of thin sections of Ti-6Al-4 V alloy due to two-sided LSP impacts.

### 3.5.2 Effect Mechanism of LSP with Different Processing Parameters on the Fatigue Property of LY2 Al Alloy

The fatigue lives of the treated samples using LSP with different processing parameters and the untreated sample are shown in Fig. 3.17 (as shown in Sect. 3.4.5). It can be clearly seen that the samples treated by the two LSP impacts on one side exhibit the best fatigue life. However, the fatigue life of the sample after two-sided LSP impact is lower than that after single-sided LSP impact.

The mechanism of the fatigue property by the single-sided LSP impact is explained by residual stress generation. The material surface is irradiated by laser shock wave. The residual stress of the impacted surface is a tensile stress state due to the propagating of laser shock wave, and the material plastic deformation occurs [51, 52]. As the shock waves propagate into the material, plastic deformation occurs to a depth at which the peak pressure no longer exceeds the Hugoniot elastic limit (HEL) of the material, and residual stresses are induced throughout the affected depth [53]. The corresponding experiment data can be found from Fig. 3.17, the affected depth of the compressive residual stresses of the first group and the second group is about 0.5 and 0.7 mm, respectively, and the leaving depth belongs to the affected depth of the tensile residual stress. Then the laser ablation stops. Because of the material reaction at the impact plane, the surrounding material is opposed to that strain with the change of volume and back to the former size. When the dynamic stresses of shock waves within a material are above the dynamic yield strength of the material, plastic deformations occurs, which continues until the peak dynamic falls below the dynamic yield strength. The plastic deformation induced by the shock waves results in strain hardening and compressive residual stresses at the material surface [54–56]. Hence, biaxial compressive residual stresses are induced.

The amplitude of the surface compressive residual stress increases with the increment of the LSP impact times which are lower than five times at best, but the affected depth of the compressive stress is almost unchangeable. Residual stresses after LSP are the stresses remaining in a metal after the shock waves are dispersed.

Such residual stresses play an important role in enhancing the fatigue properties of metallic materials [57].

Figure 3.19 shows the model of residual stress induced by LSP along the depth direction, this model was set up by David et al. [58]. With the increasing of the depth, the value of compressive residual stresses decrease until to zero, then the compressive residual stresses are changed into the tensile residual stresses in order to balance the mechanical system in the material inner and keep the whole material stable.

It can be seen from Fig. 3.19 that the supper surface is treated by LSP, and then a compressive layer to a depth of h is induced in a sheet of thickness, H. A force, F, induced by shocking a face, is assumed to act parallel to the extreme surface. And the stress acts approximately at some distance, (H/2–h/2), from the centerline of the sheet. The stress must be balanced by an equal and opposite stress of F, acting as the tensile residual stress below the shocked surface. This force comes from a uniform level of tensile stress. And the balancing tensile force acts on the opposite side of the centerline to the compressive residual stress. The experiment result is similar to this model, in Fig. 3.18 the residual stress value of the first group and the second group is negative at the depth of 0.5 and 0.7 mm or so, then that becomes positive to balance the high compressive stress value up to a deep depth.

Figure 3.20 shows the schematic diagram of two-sided LSP impact at the same time [59]. The surface and bottom of the thin sample with thickness of 2 mm are shocked by the two laser beams simultaneously. So there are two shock waves propagating from the opposite directions simultaneously. But after a period of propagating time, both of them may encounter and some parts of the waves will cancel out each other due to the simultaneous work of the two opposite shock waves. This causes the value of surface compressive residual stress decreases and the value of tensile residual stresses at the mid–plane increases sharply. This phenomenon almost occurs in the experiment. As seen in Fig. 3.18, the value of the compressive residual stress layer of the third group is lower than that of the second group although the total times of LSP impacts are two. In addition, the affected depth of the compressive residual stress layer of the third group is the shallowest among the three groups. While the value of the tensile residual stress of the third group within the material is the largest of all others, and increases sharply than the value of the compressive residual stress itself near the surface. So the capability to restrain the surface cracks initiation and propagation is cut down. At the same time, the increasing tensile residual stresses in the mid-plane cause the inner crack propagation. As a result, the fatigue property of the sample by

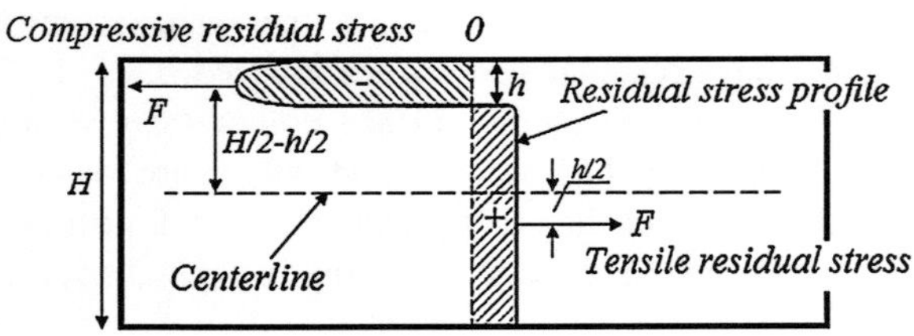


**Fig. 3.19** Model of residual stresses induced by LSP along the direction of the depth

two-sided LSP impact is lower than that by single-sided LSP impact, as shown in Fig. 3.17.

The reason why the fatigue life by two-sided LSP decreases is explained by the simplified model as shown in Fig. 3.21. It is found that the tensile residual stress in the mid-plane increases because of the counterbalance of the shock waves from the opposite directions. The experimental data are gained as enumerated in Fig. 3.17, the affected range of the tensile residual stress of the third group is wide from the depth of 0.4–0.9 mm approximately, at the same time the value of the tensile residual stress of the third group is large comparatively. In addition, the samples with the thickness of 2 mm belong to the thin sheet, and there is a certain disadvantageous counterbalance of the tensile residual stress between the frontal and reversed shocks.

Figure 3.22 shows the simulated residual stresses along the depth direction under different processing parameters by ABAQUS. With the increment of the LSP impact time, the maximum value of the compressive stress increases. The tensile residual stresses almost appear at the mid-plane of the samples, and the value of tensile residual stress of the third group is the highest among the three groups. From this picture, it can be also found that the compressive residual stresses of the third group occur in the mid-plane. The similar result is obtained by the experiment data in Fig. 3.18.

If the laser beam is split into two beams with the same energy by the semi-permeable semi-anti-optical lens with the angle of 45°, and the two beams are simultaneously focused onto both sides of the section with a thicker thickness, the enhancement effects of LSP with two sides will be better, which needs more systematic investigations.

## 3.6 Summary

In this chapter, the effects of strain rate on the fracture morphology and tensile properties of LY2 Al alloy were systematically investigated, and the effect of different LSP processing parameters on the fatigue property of LY2 Al alloy were also studied. Special attention was paid to the fatigue striations of the non-LSPed and LSPed samples. Furthermore, the relation between the UTS and strain rate

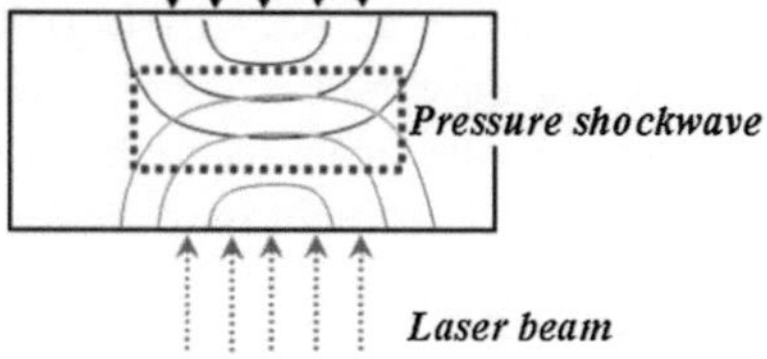


**Fig. 3.20** Schematic of two-sided laser shock processing at the same time

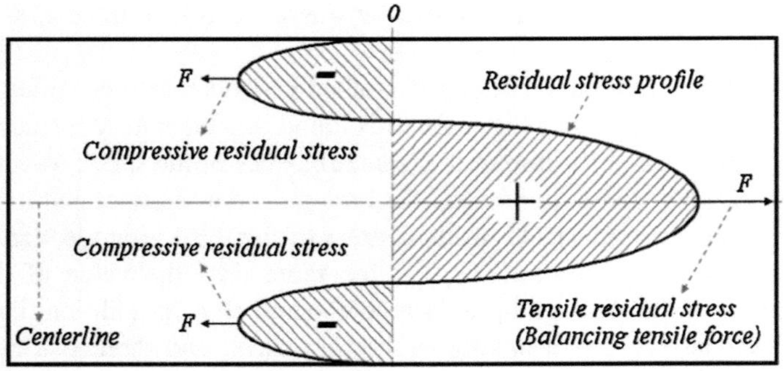


**Fig. 3.21** Model of residual stresses induced by two-sided LSP along the direction of the depth

**Fig. 3.22** Simulation of the residual stresses along the depth direction under different technologic parameters

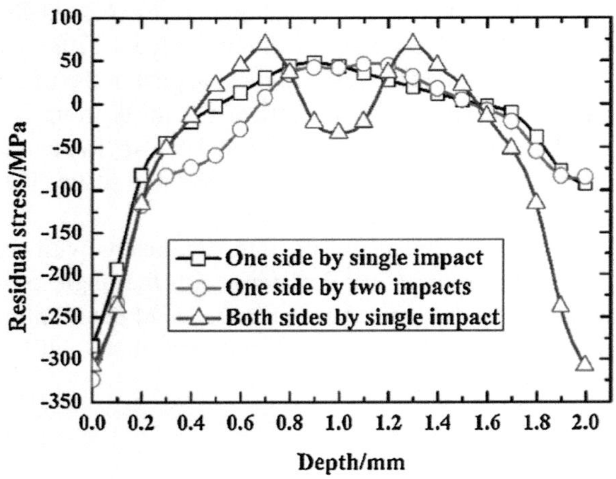


sensitivity was also addressed. Some important conclusions have been obtained as follows:

1. LSP can remarkably improve the UTS, the flow stress and the elongation of the tensile sample manufactured by LY2 Al alloy. The grain in the shocked region subjected to LSP was clearly refined, and the grain size is about 100–200 nm at the top surface, which has an obvious attribution to the improvement of UTS and elongation the tensile sample manufactured by LY2 Al alloy.
2. The strain rate plays an important role on the UTS (or flow stress) of the treated sample manufactured by LY2 Al alloy. With the increment of the strain rate, the elongation decreases gradually, and there is a slight decrease in UTS at the strain rate from 0.00001 to 0.0001 $s^{-1}$, while there is an obvious increase in UTS at the strain rate from 0.0001 to 0.1 $s^{-1}$. The phenomenon of DSA was found in the treated sample by LSP at the strain rate ranging from 0.00001 to 0.0001 $s^{-1}$, which may be explained by the dynamic interaction between

diffusing solute atom and the mobile dislocation in the SPD layer of LY2 Al alloy at the strain rate ranging from 0.00001 to 0.0001 $s^{-1}$.

3. The typical dimple morphology appears on the fracture surfaces of LY2 Al alloy subjected to LSP at all strain-rates. When the strain-rate increased from 0.001 to 0.1 $s^{-1}$, there was the dimple coalescence morphology on the fracture surfaces, and the dimple becomes gradually larger and deeper. Fracture morphology suggests the possibility that LY2 Al alloy evolves towards a more ductile dimple fracture mode associated with dimple coalescence at the higher strain-rates.
4. LSP is considered as one of the most promising techniques in terms of its ability to induce compressive residual stresses, which can enhance the fatigue performance of the materials. The effect of processing parameters used in the LSP on the residual stress and micro-hardness at the surface and the substrate of the edge of the turbojet engine blade manufactured by LY2 Al alloy and the LCF performance of the notched samples manufactured by LY2 Al alloy are experimentally investigated. Some important conclusions can be obtained. The thickness of the plastic deformation layer near the surface generated by the shockwave in LSP is higher than 2.0 mm. With increasing the number of the laser impacts in LSP, the compressive residual stresses and the micro-hardness at the surface of the sample increases obviously. However, the residual stresses and the micro-hardness at the locations far away from the surface (i.e. $\geq$ a few mm) are almost not influenced by the number of the laser impacts. The micro-indents at the surface of the sample will be generated after LSP. Moreover, by comparing with the untreated sample, the LCF performance of the sample treated by LSP is obviously increased, which may be due to the high compressive residual stresses and micro-hardness at the treated surface.
5. Compressive residual stress layer is generated at the surface layer of LY2 Al alloy. The fatigue life of the samples by two LSP impacts at one side are higher than that those by one laser impact. The fatigue lives of the samples by two-sided LSP are lower than those by single-sided LSP, but higher than those of the untreated samples. Fatigue property of the metal component depends on the distribution of the residual stress. The inner tensile residual stresses at the mid-plane of the treated samples increase and their fatigue lives decrease due mainly to simultaneous work of the two opposite shock waves during two-sided LSP impacts.
6. The LCF striations of LSPed samples are characterized as a non-crystallography brittle fracture, and the cloud-like extruded ridges are the important microscopic features. The LSPed samples render the effect of cyclic hardening in the process of fatigue cracks propagation, which effectively hinders the occurrence of secondly cracks, therefore improves the fatigue performance.

## References


1. Yang, C. H., Hodgson, P. D., & Liu, Q. C. (2008). Geometrical effects on residual stresses in 7050–T7451 Al alloy rods subject to laser shock peening. *Journal of Materials Processing Technology, 201*, 303–309.
2. Zhang, H., & Yu, C. Y. (1998). Laser shock processing of 2024–T62 Al alloy. *Materials Science and Engineering A, 257*, 322–327.
3. Montross, C. S., Ye, L., Wei, T., Clark, G., & Mai, Y. W. (2002). Laser shock processing and its effects on microstructure and properties of metal alloys: A review. *International Journal of Fatigue, 24*, 1021–1036.
4. Zhang, Y. K., Hu, C. L., Cai, L., Yang, J. C., & Zhang, X. R. (2001). Mechanism of improvement on fatigue life of metal by laser–excited shock waves. *Applied Physics A, 72*(2), 113–116.
5. Rubio–González, C., Ocaña, J. L., & Gomez–Rosas, G. (2004). Effect of laser shock processing on fatigue crack growth and fracture toughness of 6061–T6 Al alloy. *Materials Science and Engineering A, 386*, 291–295.
6. Zhang, X. C., Xu, B. S., & Wang, H. D. (2007). Residual stress distributions within high–temperature coatings. *Surface and Coatings Technology, 201*, 6660–6662.
7. Zhang, X. C., Xu, B. S., & Wang, H. D. (2008). Effect of graded interlayer on the mode I edge delamination by residual stresses in multilayer coating–based systems. *Applied Surface Science, 254*, 1881–1889.
8. Hu, Y. X., & Yao, Z. Q. (2008). Overlapping rate effect on laser shock processing of 1045 steel by small spots with Nd:YAG pulsed laser. *Surface and Coatings Technology, 202*, 1517–1525.
9. Lavender, C. A., Hong, S. T., Smith, M. T., Johnson, R. T., & Lahrmanc, D. (2008). The effect of laser shock peening on the life and failure mode of a cold pilger die. *Journal of Materials Processing Technology, 204*, 486–491.
10. Dane, C. B., Hackel, L. A., & Daly, J. (1997). Laser peening of metals–enabling laser technology. *Advanced Material Processing, 5*, 13–27.
11. Vaccari, J. A. (1992). Laser shocking extends fatigue life. *American Machine, 6*, 62–64.
12. Garcia–Bernal, M. A., Mishra, R. S., Verma, R., & Hernandez–Silva, D. (2009). High strain rate superplasticity in continuous cast Al–Mg alloys prepared via friction stir processing. *Scripta. Materialia, 60*(10), 850–853.
13. Garcia–Infanta, J. M., Zhilyaev, A. P., Sharafutdinov, A., Ruano, O. A., & Carreno, F. (2009). An evidence of high strain rate superplasticity at intermediate homologous temperatures in an Al–Zn–Mg–Cu alloy processed by high–pressure torsion. *Journal of Alloys and Compounds, 473*, 163–166.
14. Boyce, B. L., & Dilmore, M. F. (2009). The dynamic tensile behavior of tough, ultrahigh–strength steels at strain–rates from 0.0002 to 200 $s^{-1}$. *International Journal of Impact Engineering, 36*, 263–271.
15. Fan, G. J., Wang, G. Y., Choo, H., Liaw, P. K., Park, Y. S., Han, B. Q., et al. (2005). *Scripta Material, 52*, 929–933.
16. Kaibyshev, R., Avtokratova, E., Apollonov, A., & Davies, R. (2006). High strain rate superplasticity in an Al–Mg–Sc–Zr alloy subjected to simple thermo mechanical processing. *Scripta Materialia, 54*, 2119–2124.
17. Han, B. Q., Huang, J. Y., Zhu, Y. T., & Lavernia, E. J. (2006). Strain rate dependence of properties of cryomilled bimodal 5083 Al alloys. *Acta Materialia, 54*, 3015–3024.
18. Kocks, U. F., & Mecking, H. (2003). Physics and phenomenology of strain hardening: The FCC case. *Progress in Materials Science, 48*(3), 171–273.
19. Kapoor, R., & Chakravartty, J. K. (2007). Deformation behavior of an ultrafine–grained Al–Mg alloy produced by equal–channel angular pressing. *Acta Materialia, 55*(16), 5408–5418.

20. Kuo, T. Y., & Lin, H. C. (2006). Effects of pulse level of Nd–YAG laser on tensile properties and formability of laser weldments in automotive Al alloys. *Materials Science and Engineering A, 416*(1–2), 281–289.
21. Liu, C., Northwood, D. O., & Bhole, S. D. (2004). Tensile fracture behavior in $CO_2$ laser beam welds of 7075–T6 Al alloy. *Materials and Design, 25*(7), 573–577.
22. Schino, A. D., & Kenny, J. M. (2003). Grain size dependence of the fatigue behaviour of a ultrafine–grained AISI 304 stainless steel. *Materials Letters, 57*(21), 3182–3185.
23. Lu, J. Z., Luo, K. Y., Zhang, Y. K., Cui, C. Y., Sun, G. F., Zhou, J. Z., et al. (2010). Grain refinement of LY2 Al alloy induced by ultra–high plastic strain during multiple laser shock processing impacts. *Acta Materialia, 58*(11), 3984–3994.
24. Sakka, Y., Suzuki, T. S., Morita, K., Nakano, K., & Hiraga, K. (2001). Colloidal processing and superplastic properties of zirconia– and alumina–based nanocomposites. *Scripta Materialia, 44*(8–9), 2075–2078.
25. Nascimento, F., Santos, T., Vilaça, P., Miranda, R. M., & Quintino, L. (2009). Microstructural modification and ductility enhancement of surfaces modified by FSP in Al alloys. *Materials Science and Engineering A, 506*(1–2), 16–22.
26. Mckenzie, P. W. J., & Lapovok, R. (2010). ECAP with back pressure for optimum strength and ductility in Al alloy 6016. Part 2: Mechanical properties and texture. *Acta Materialia., 58*(9), 3212–3222.
27. Csontos, A. A., & Starke, E. A. (2005). The effect of inhomogeneous plastic deformation on the ductility and fracture behavior of age hard enable Al alloys. *International Journal of Plasticity, 21*(6), 1097–1118.
28. Gu, C. D., Lian, J. S., Jiang, Q., & Jiang, Z. H. (2007). Ductile–brittle–ductile transition in an electrodeposited 13 nanometer grain sized Ni–8.6 wt % Co alloy. *Materials Science and Engineering A, 459*, 75–81.
29. Han, B. Q., Huang, J. Y., Zhu, Y. T., & Lavernia, E. J. (2006). Effect of strain rate on the ductility of a nanostructured Al alloy. *Scripta Materialia, 541*(6), 175–1180.
30. Dudrová, E., & Kabátová, M. (2008). Fractography of sintered iron and steels. *Powder Metallurgy Progress, 8*(2), 59–75.
31. Arpan, D., & Soumitra, T. (2008). Geometry of dimples and its correlation with mechanical properties in austenitic stainless steel. *Scripta Materials, 59*(9), 1014–1017.
32. See, D. W., Dulaney, J. L., & Clauer, A. H. (2002). The air force manufacturing technology laser peening initiative. *Surface Engineering, 18*, 32–36.
33. King, A., Steuwer, A., & Woodward, C. (2006). Effects of fatigue and fretting on residual stresses introduced by laser shock peening. *Materials Science and Engineering A, 12*, 435–436.
34. King, A., Evans, A. D., & Withers, P. J. (2005). The effect of fatigue on residual peening stresses in aerospace components. *Materials Science Forum, 490–491*, 340–345.
35. Prevey, P., & Jayaraman, N. (2005). A design methodology to take credit for residual stresses in fatigue limited designs. *Journal of ASTM International, 2*(8), 61–76.
36. Larsen, J. M., Worth, B. D., & Annis, C. G. (1996). An assessment of the role of near–threshold crack growth in high–cycle–fatigue life prediction of aerospace titanium alloys under turbine engine spectra: High cycle fatigue. *International Journal of Fracture, 80*, 237–255.
37. Duo, P., Liu, J., & Dini, D. (2007). Evaluation and analysis of residual stresses due to foreign object damage. *Mechanics of Materials, 39*, 199–211.
38. Nowell, D., Duo, P., & Stewart, I. F. (2003). Prediction of fatigue performance in gas turbine blades after foreign object damage. *International Journal of Fatigue, 25*, 963–969.
39. Hu, Y. X., Yao, Z. Q., & Hu, J. (2006). 3–D FEM simulation of laser shock processing. *Surface and Coatings Technology, 201*, 1426–1435.
40. Masse, J. E., & Barreau, G. (1995). Surface modification by laser induced shock waves. *Surface Engineering, 11*, 131–142.

41. Clauer, A. H., Holbrook, J. H., & Fairand, B. P. (1981). Effects of laser induced shock waves on metals. In M. A. Meyers & L. E. Murr (Eds.), *Shock waves and high–strain–rate phenomena in metals* (pp. 675–702). New York: Plenum Publishing Corporation.
42. Fourier, J. (1990). Mechanical effects induced by shock waves generated by high–energy laser pulses. *Material Manufacturing Processing, 5*, 144–147.
43. Zhang, Z. P., Li, J., & Sun, Q. (2009). Two parameters describing cyclic hardening/softening behaviors of metallic materials. *Journal of Materials Engineering and Performance, 18*(3), 237–244.
44. Scherpereel, X., Peyre, P., & Fabbro, R. (2005). Modifications of mechanical and electrochemical properties of stainless steels surfaces by laser shock processing. *SPIE, 3097*, 546–557.
45. Peyre, P., Fabbro, R., Merrien, P., & Lieurade, H. P. (1996). Laser shock processing of Al alloys. Application to high cycle fatigue behaviour. *Materials Science and Engineering A, 210*, 102–113.
46. Chien, W. Y., Pan, J., Close, D., & Ho, S. (2005). Fatigue analysis of crankshaft sections under bending with consideration of residual stresses. *International Journal of Fatigue, 27*, 1–19.
47. Farrahi, G. H., Lebrun, J. L., & Couratin, D. (1995). Effect of shot peening on residual stress and fatigue life of spring steel. *Fatigue and Fracture of Engineering Materials and Structures, 18*, 211–220.
48. Guagliano, M., & Vergani, L. (2004). An approach for prediction of fatigue strength of shot peened components. *Engineering Fracture Mechanics, 71*, 501–512.
49. Wang, S. P., Li, Y. J., Yao, M., & Wang, R. Z. (1998). Compressive residual stress introduced by shot peening. *Journal of Materials Processing Technology, 73*, 64–73.
50. Ding, K., & Ye, L. (2003). FEM simulation of two sided laser shock peening of thin sections of Ti–6Al–4 V alloy. *Surface Engineering, 19*, 127–133.
51. Huang, S., Zhou, J. Z., Jiang, S. Q., Zhu, Y. B., & Hu, L. L. (2010). Dynamical analysis on stresses in metal caused by laser shot peening. *Chinese Journal Lasers, 37*, 256–260. (in Chinese).
52. Hu, Y. X., Gong, C. M., Yao, Z. Q., & Hu, J. (2009). Investigation on the non–homogeneity of residual stress field induced by laser shock peening. *Surface and Coatings Technology, 203*, 3503–3508.
53. Peyre, P., Berthe, L., Scherpereel, X., & Fabbro, R. (1998). Laser–shock processing of Al–coated 55C1steel in water–confinement regime, characterization and application to high–cycle fatigue behavior. *Journal of Materials Science, 33*, 1421–1429.
54. Arif, A. F. M. (2003). Numerical prediction of plastic deformation and residual stresses induced by laser shock processing. *Journal of Materials Processing Technology, 136*, 120–138.
55. Peyre, P., & Fabbro, R. (1995). Laser shock processing: A review of the physics and applications. *Optical and Quantum Electronics, 27*, 1213–1229.
56. Ocaña, J. L., Morales, M., Molpeceres, C., & Torres, T. (2004). Numerical simulation of surface deformation and residual stresses fields in laser shock processing experiments. *Applied Surface Science, 238*, 242–248.
57. Han, B., & Ju, D. Y. (2009). Compressive residual stress induced by water cavitation peening: A finite element analysis. *Materials and Design, 30*, 3325–3332.
58. Kirk, D. (2003). Relationships between Almen strip thickness and arc heights. *The Shot Peener, 17*, 22–27.
59. Lu, J. Z. (2010). Investigation of laser shock processing on the mechanical properties and micro-plastic deformation mechanism of LY2 aluminum alloy. Dissertation, Jiangsu University. (in Chinese).

# Chapter 4
# Grain Refinement of LY2 Al Alloys Induced by Multiple Laser Shock Processing Impacts

**Abstract** This chapter presents the effects of the multiple LSP impacts on the residual stresses in the depth direction, focusing especially on the change of microstructure in different layers in the treated sample by multiple LSP impacts, and clearly reveals the underlying refinement mechanism of plastic deformation after multiple LSP impacts on LY2 Al alloy.

## 4.1 Introduction

Surface conditions and their modifications during laser irradiation play important roles in laser-matter interaction, which has important applications, such as micromachining, thin films deposition, biological investigations, photolithography, magnetic materials, etc. LSP is a new and promising surface treatment technique to improve the fatigue durability, corrosion, wear resistance and other mechanical properties of the metals and alloys, during which the generated shock wave can introduce a deep compressive residual stresses into the metallic materials [1–6].

Al alloy is widely used in aerospace and automotive industries due to its light weight and moderate strength. But its moderate strength and corrosion resistance limit its wide use. Many researchers has devoted to the study of the modification of the Al alloy, such as shot peening [7], cold rolling [8], ball milling [9], surface mechanical attrition treatment [10] and LSP [1–6]. Among these technologies, in particular, LSP, as one of the effective ways to improve the surface mechanical properties of Al alloys, has been intensively studied by many researchers [11–18]. For example, the geometrical effects on residual stresses in 7050-T7451 Al alloy rods subjected to LSP was investigated [11], and the effects of a single LSP on residual stress distribution in 7085-T7651 Al alloy [12] was also exhibited. Besides, the relationship between the principal residual stresses and the depth of the 6061-T6 Al sample by LSP was established [1, 13]. A number of studies [3, 6, 14, 15] have investigated the high level compressive residual stresses produced in

Y. Zhang et al., *Laser Shock Processing of FCC Metals*,
Springer Series in Materials Science 179, DOI: 10.1007/978-3-642-35674-2_4,

Al alloys during the LSP as well as the improvement of fatigue life made by the compressive stress magnitude. However, most of the above researches have focused on the residual stress distribution at the surface and in depth direction after a single LSP. In fact, multiple LSP impacts may be the more effective method to significantly increase the compressive residual stress in depth direction. The influence of the LSP with repeated impacts on the residual stress distribution of 7,075 Al alloy was investigated [16], and the affected layer of the 4,340 steel sheet by LSP was analyzed and simulated by experiment [17] and finite element method [18]. The results show that the improvement of residual stress is directly related to the generation of the dislocation and the micro-structural deformation near the surface during LSP process. However, the physical mechanism during LSP on the residual stress improvement lags well behind, and the micro-structural deformation mechanism during multiple LSP impacts is very important in this respect. At the same time, the interaction between shock waves and the resultant micro-structural changes during multiple LSP impacts is still pending, and the enhancement mechanism of multiple LSP impacts on Al alloys is worth investigation.

This chapter investigates the effects of the multiple LSP impacts on the residual stresses in the depth direction, focusing especially on the change of micro-structure evolution in different layers in the treated sample by multiple LSP impacts. The underlying refinement mechanism of plastic deformation after multiple LSP impacts on LY2 Al alloy is clearly revealed.

## 4.2 Micro-Structural Characterization

The samples manufactured by LY2 Al alloy were cut into a cuboid shape with dimensions of $30 \times 10 \times 3\ mm^3$ (width × length × thickness). The chemical composition and mechanical properties of LY2 Al alloy were shown in Tables 2.3 and 2.4, respectively. Prior to the LSP treatment, the sample surfaces were polished with SiC paper with different grades of roughness (from 500 to 2,400), followed by cleaning in deionized water. Ultrasound in ethanol was used to degrease the sample surface, and LSP experiments were conducted shortly after preparation.

The LSP experiments were performed using a Q-switched Nd: glass laser operating at 1 Hz repetition-rate with a wavelength of 1,054 nm and the FWHM of the pulses was about 20 ns. The spot diameter was 3 mm. Samples were submerged into a water bath when they were processed by LSP. A water layer with a thickness of about 1 mm was used as the transparent confining layer and the professional Al tape with a thickness of 100 μm was used as an absorbing layer to protect the sample surface from thermal effect. The Al tape was replaced after each impact during multiple LSP impacts. Laser energy was 5 J. The processing parameters used in LSP were shown in Table 4.1 in detail. The treated area at the sample surface was $10 \times 10\ mm^2$, and a five-axis motion system was used to generate the pulse.

**Table 4.1** The processing parameters used in LSP

| Type | Value |
|---|---|
| Beam div. of output (mrad) | $\leq$0.5 |
| Spot diameter (mm) | 3 |
| Pulse energy (J) | 5 |
| Repetition-rate (Hz) | 1 |
| Laser wavelength (nm) | 1,054 |
| Export stability | $\leq \pm 5$ % |

The residual stresses after multiple LSP impacts were determined by using XRD with $\sin^2 \psi$ method. Prior to the measurement of the residual stress along the depth direction, the electropolishing material removal method was used. The X-ray beam diameter was about 2 mm. The X-ray source was Cr-K$\alpha$ ray and the diffraction plane was $\alpha$ phase (420) plane. The feed angle of the ladder scanning was 0.1° per second. The scanning starting angle and terminating angle were 159° and 168°, respectively. The measurements were repeated five times for each condition, and an average value was used.

The samples used for metallographic observation as a function of the impact time were cut and made a section perpendicular to the sample surface, and then subjected to several successive steps of grinding and polishing. After that, the vertical sections were etched using a Keller's reagent that consists of 190 ml of $H_2O$, 5 ml of $HNO_3$, 2 ml of HF, and 3 ml of HCl for 10 min at room temperature, and then characterized by a scanning electron microscopy (SEM) and cross-sectional optical microscopy (OM).

The microstructural evolution of the different layers in the treated samples subjected to multiple LSP impacts was characterized by using a JEM-2100 transmission electron microscope (TEM) operated at a voltage of 200 kV. TEM foils from a given depth below the surface were prepared by a combination of single and twin-jet electropolishing. All micrographs presented are bright-field images recorded under two-beam conditions.

## 4.3 Residual Stress Distribution of the Hardening Layer

The residual stress profiles of the treated samples after multiple LSP impacts with different impact time as functions of the distance from the top surface are shown in Fig. 4.1. The substrates are approximately in the zero-stress state, indicating that the effect of initial residual stress on the shock waves may be ignored [19]. It can be noted from Fig. 4.1 that the significant compressive residual stresses mainly exist in near-surface regions for all cases and the top surfaces have the maximum values of compressive residual stresses.

The peak surface compressive residual stress for a single LSP impact and two LSP impacts is 81 and 102 MPa, and the corresponding depth of compressive

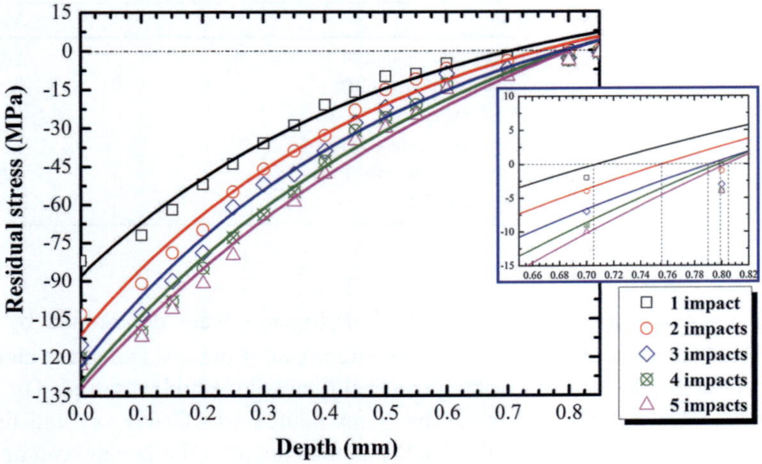


**Fig. 4.1** Residual stress profiles of the hardening layer after multiple LSP impacts as a function of the impact time

residual stress is 0.707 and 0.756 mm, respectively. The peak surface compressive residual stress and the depth of compressive residual stress are significantly increased to 116 MPa and to 0.79 mm, respectively, as a result of three LSP impacts on the sample surface. After four LSP impacts, the peak value of surface compressive residual stress is increased to 123 MPa, and the depth of compressive residual stress reaches about 0.80 mm. It can be seen that the surface compressive residual stress is increased by 25.93 and 13.73 % when the impact time increases from one to two and from two to three, whereas the surface compressive residual stress is increased by 6.89 % when the impact time increases from three to four, but the surface compressive residual stress is kept to about 123 MPa after the multiple LSP with four and five LSP impacts.

From the above discussion, it can be seen from Fig. 4.2 that the increasing rate of surface compressive residual stress decreases almost linearly with the impact time, but the increase of surface residual stresses gradually reaches the saturated state when the impact time exceeds four. The similar results can be seen elsewhere [17, 18]. From Fig. 4.3, it can be seen that the depth of compressive residual stress gradually increases with increasing impact time, and the depth of compressive residual stress keeps a constant value when the impact time exceeds four.

Although there is some discrepancy between the proper values and the measured values which may be due to that the magnitudes of the residual stress in depth direction are measured by using electropolishing material removal method, but we can also obtain the changing trend of residual stress distribution in depth direction as functions of the distance from the top surface. It is well known that cumulative impacts have a very superficial effect on the residual stress levels in the superficial layers [16], and the increasing rate of the residual stresses decreases as the impact time increases [17, 18]. Hence, it can be concluded that the lower the distance from the top surface, the more important the influence of the impact time

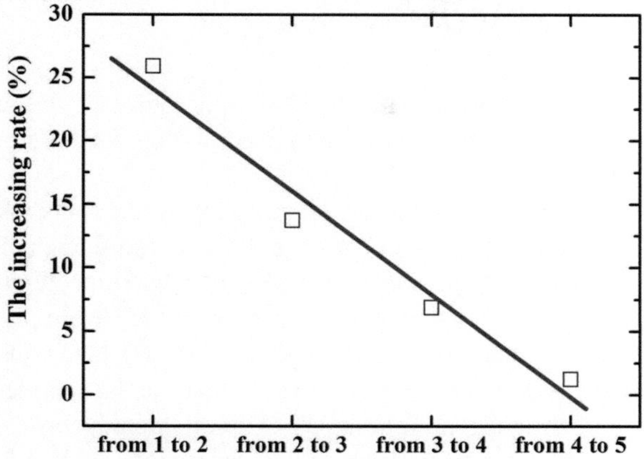


**Fig. 4.2** The comparison between the increasing rate of surface residual stress and the impact time

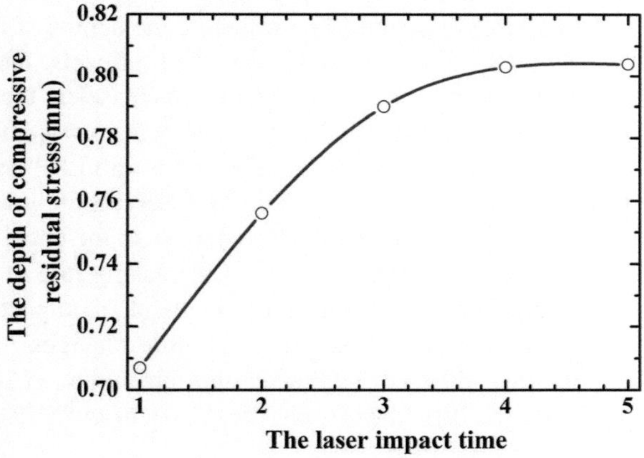


**Fig. 4.3** The comparison between the affected depth of compressive residual stress and the impact time

on the residual stress, but when the impact time exceeds four, the magnitude of surface residual stress and the affected depth are almost not influenced by the impact time.

## 4.4 Plastic Deformation and Microstructural Feature

As the strain decreases from maximum at the top surface layer to zero in the substrate, the structure evolution process during multiple LSP impacts may be signed by the microstructure characteristics (with different strains) at different depths. In order to systematically understand the microstructures developed in the hardening layer of the samples during multiple LSP impacts, detailed cross-sectional SEM observations as a function of the impact time are needed. For the sake of convenience, the cross-sections of the treated samples after three LSP impacts are taken as samples for observations by OM and TEM.

### 4.4.1 SEM Observations

Figure 4.4 presents the SEM morphologies of the cross-sections of the samples with and without LSP immersed in the Keller's reagent for 10 min at room temperature.

Figure 4.4a is the SEM morphology of the cross-section without LSP, and Figs. 4.4b, c, d, e and f are the SEM morphologies of the cross-sections with one, two, three, four, and five LSP impacts, respectively. In this study, the plastic deformation (PD) layer is composed of the severe plastic deformation (SPD) layer and the minor plastic deformation (MPD) layer. Obviously, after multiple LSP impacts, the thicknesses of the PD layer in the sample surfaces are uniform, but the microstructure morphologies of the hardening layer with different impacts differ from those in the corresponding substrates. Table 4.2 shows that the depths of the PD layers are 372, 467, 506, 522, and 529 μm and the depths of the SPD layers are 239, 361, 448, 494, and 511 μm, while the depths of the MPD layers are 133, 106, 58, 28 and 18 μm, with one to five LSP impacts, respectively.

It can be seen in Table 4.2 that the increases in the depths of the SPD layer and the PD layer after the first impact are remarkable. Between the second and third LSP impacts and between the third and fourth LSP impacts, the increasing rates of the depths of the SPD layers and the PD layers are low. But between the fourth and fifth LSP impacts, there are little changes in the corresponding depth values. It can be observed from Fig. 4.4 that the increasing rate of the affected depth in plastic deformation decreases with the increment of impact time, and when the impact time is from four to five, there are little changes in the depth of plastic deformation. In additional, similar results are obtained in the decreasing rate of the depths for the MPD layers, but the depths of the MPD layers gradually decrease as the impact time increases.

It can be seen from Fig. 4.5 that the depth of the compressive residual stress layer is obviously deeper than that of plastic deformation layer with the same LSP impact time. In general, the depth of the compressive residual stress layer does not correspond to that of the PD layer. Usually, the former is larger than the latter. However, it is clear that their changing trends as functions of the LSP impact time are similar.

### 4.4.2 OM Observations

Samples subjected to three LSP impacts for the observation of the cross-sectional microstructures are immersed in the Keller's reagent for 10 min at room temperature by OM. Figure 4.6a, b and c show the typical magnified optical morphologies of the cross-section images obtained from the SPD layer, the MPD layer and the substrate, respectively. It can be clearly seen from Fig. 4.6 that the average grain size in the SPD layer is about 3–5 μm, while that in the substrate is larger

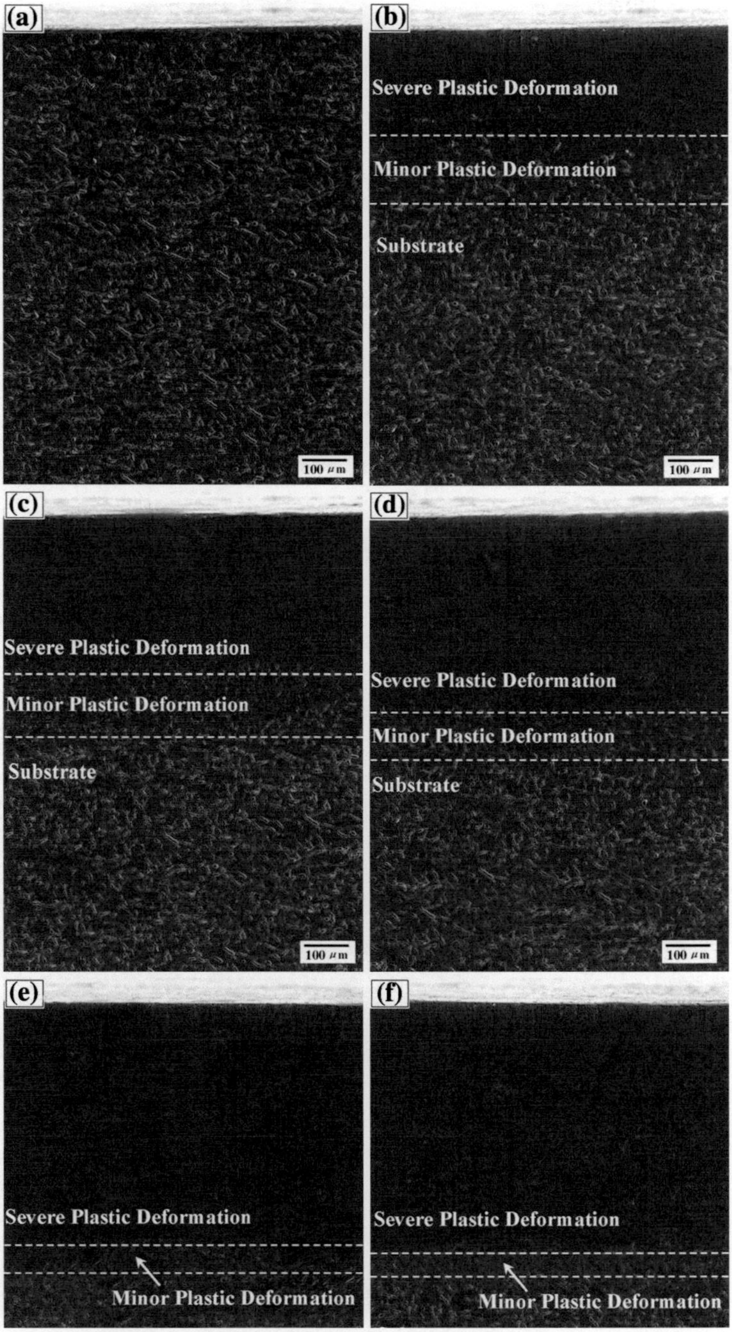


**Fig. 4.4** SEM morphologies of the cross-sections immersed in the Keller's reagent for 10 min at room temperature

**Table 4.2** The depths of different layers as a function of LSP impact time

| The depth of (μm) | 1 LSP impact | 2 LSP impacts | 3 LSP impacts | 4 LSP impacts | 5 LSP impacts |
|---|---|---|---|---|---|
| Grind-hardened layer | 372 | 467 | 506 | 522 | 529 |
| SPD layer | 239 | 361 | 448 | 494 | 511 |
| MPD layer | 133 | 106 | 58 | 28 | 18 |

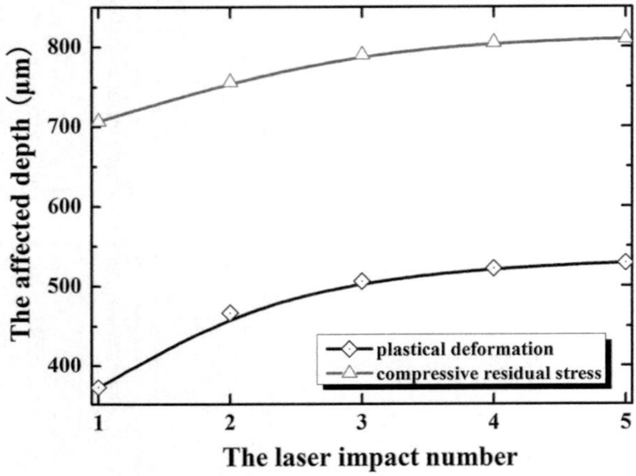


**Fig. 4.5** The comparison of the affected depths between the plastic deformation and compressive residual stress

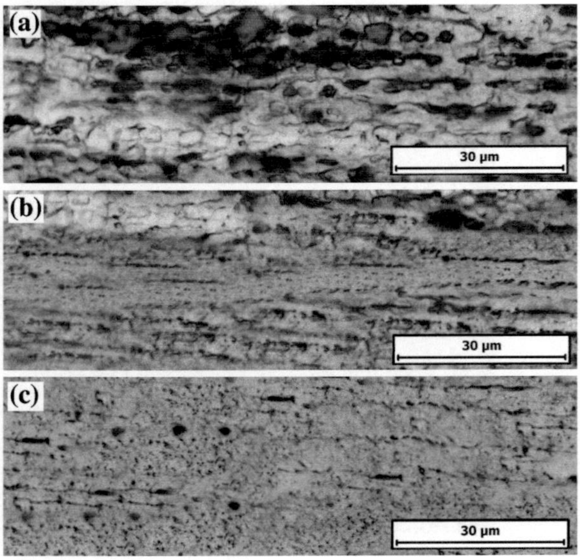


**Fig. 4.6** Optical cross-sections images of **a** the SPD layer, **b** the MPD layer, and **c** substrate subjected to three LSP impacts immersed in the Keller's reagent for 10 min at room temperature, showing the grain structures

than 10 μm. The grain size in the SPD layer is smaller compared to that in substrate after three LSP impacts. Furthermore, the grain is more homogenous compared to that in the substrate.

The relation between the grain size and the fatigue behavior of a ultrafine-grained AISI 304 SS indicated a strong improvement in both the tensile and fatigue resistance by grain refinement [20, 21]. Consequently, it can be concluded that the grain in the SPD layer is clearly refined after multiple LSP impacts, which is favorable to the improvement on the fatigue life of LY2 Al alloy.

### *4.4.3 TEM Observations*

Figure 4.7 shows the TEM observations in the SPD layer, the MPD layer and the substrate subjected to three LSP impacts, in which three typical deformation-induced microstructure features are identified: dislocation lines (DLs), dislocation tangles (DTs) and dense dislocation walls (DDWs).

Figure 4.7a and b show the TEM images of grains in the substrate at the depth of 480 μm from the top surface, it can be clearly seen from which that the size of the original grain is about 10–12 μm. Figure 4.7c shows the TEM image in the MPD layer at the depth of 380 μm from the top surface, and it can be seen that there are plenty of DLs in the MPD layer. Figure 4.7d and e show the TEM images in the SPD layer at the depth of 200 and 100 μm from the top surface. In some grains, the dislocation distribution is not uniform and in some regions the dislocation density is pretty high. High-density DTs are randomly arranged in some grains, as can be seen from Fig. 4.7d. DDWs are frequently seen inside some grains, as shown in Fig. 4.7e. DDWs are believed to result from dislocation accumulation and rearrangement for minimizing the total energy state. Figure 4.7f shows the TEM images of the top surface, and it can be clearly seen that the original coarse grains are subdivided into many subgrains or refined grains with a width of 3–5 μm. Figure 4.7g and h are the higher magnifications of the ellipses [A] and [B] in Fig. 4.7f, respectively. It can be seen that there are plenty of DLs inside subgrains, and the pile-up of DLs contributes to DTs and DDWs in the top surface from Fig. 4.7h, and eventually the pile-up of dislocation structure contributes to the formation of the subgrain boundaries in Fig. 4.7g. In additional, it can be seen in ellipse C from Fig. 4.7f that the grain size is about 100–200 nm.

After three LSP impacts, it can be seen that the average grain size in the substrate is larger than 10 μm from Fig. 4.7a and b, and that in the top surface is about 3 μm from Fig. 4.7f, which is consistent with the OM results in the Sect. 4.4.2. From the above experimental results, it can be concluded that after three LSP impacts there are plenty of DLs in the MPD layer, and plenty of DDWs and DTs in the SPD layer, indicating the change of dislocation structures with a decrease of depth from the top surface.

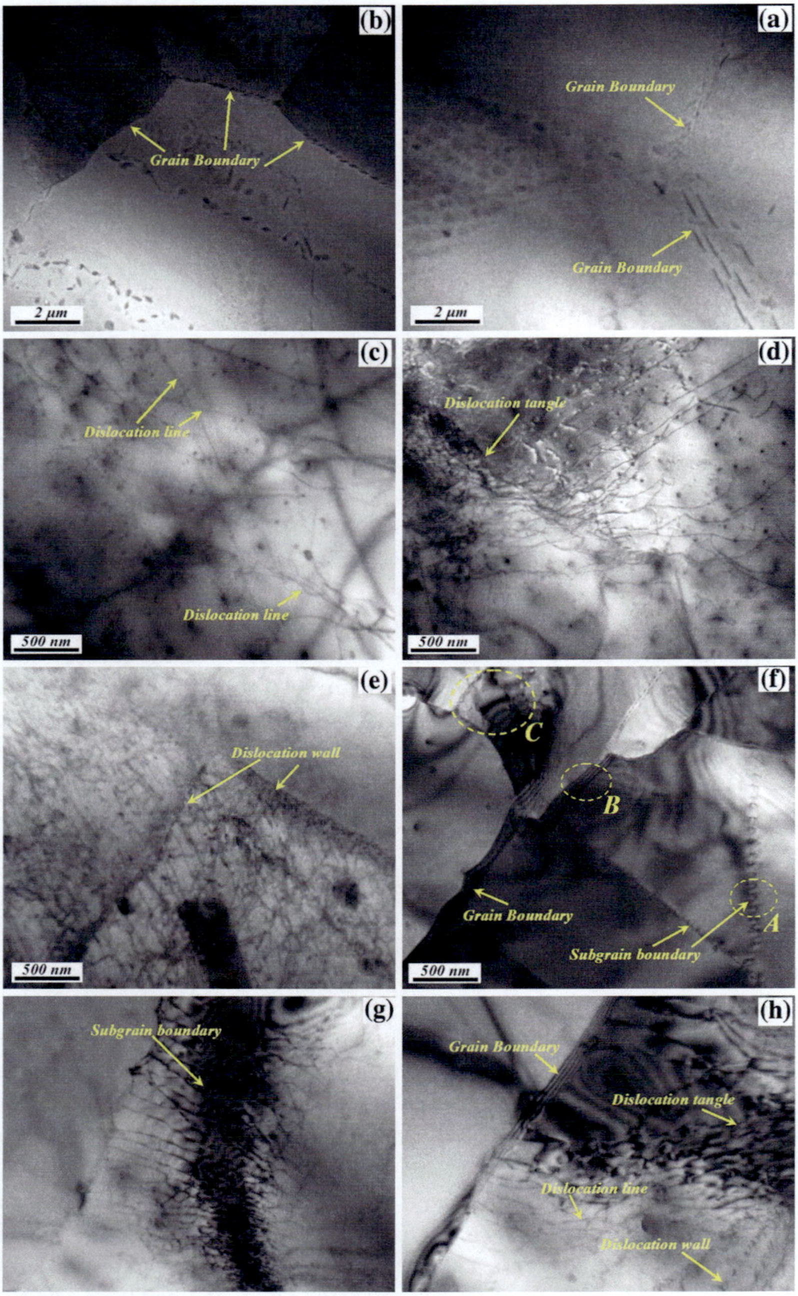


**Fig. 4.7** Typical TEM images of different layers after three LSP impacts. **a** and **b** in the substrate at the depth of 480 μm from the *top* surface, **c** in the MPD layer at the depth of 380 μm from the *top* surface, **d** and **e** in the SPD layer at the depth of 200 and 100 μm from the *top* surface, respectively, **f** at the *top* surface in the SPD layer, **g** and **h** the higher magnifications of the *ellipses* *A* and *B* in Fig. 4.7f

## 4.5 Micro-Structural Strengthening Mechanism

### *4.5.1 Enhancement Mechanism of Multiple LSP Impacts*

In the previous LSP studies [2–7, 12, 14, 16, 19, 22], the compressive residual stresses are emphasized since these are the most significant contribution to the improvement of fatigue properties. The presence of compressive residual stresses is a result of the shock pressure exerted on the sample surface during multiple LSP impacts. Furthermore, it is well known that residual stresses in metal materials are often the result of micro-plastic deformation accompanying the microstructure changes [23]. As a result, it is reasonable to assume that the LSP induced strengthening in metal materials is due to the generation of dislocations.

The schematic illustrations of the microstructure characteristics of the hardening layer subjected to three LSP impacts are shown in Fig. 4.8. After three LSP impacts, the change of dislocation structure can be also clearly seen at different layers, i.e., it varies from DLs to DTs and DDWs, to subgrains or refined grains as functions of the distance from the top surface. After multiple LSP impacts, the grains in the SPD layer are clearly refined and there are plenty of DLs and DTs with high density in the SPD layer. As a result of the grain refinement, the shocked area is strengthened according to the classical dislocation theory [24], where

$$\tau_N = \frac{2\alpha\mu b_N}{D} \tag{4.1}$$

and

$$\tau_P = \frac{2\alpha\mu b_P}{D} + \frac{\gamma}{b_P} \tag{4.2}$$

here $\mu$ is the shear modulus ($\sim$35 GPa for Al alloy), $\gamma$ is the stacking fault energy (104–142 mJ/m$^2$ for Al alloy [25]), D is the grain size, and $b_N$ and $b_P$ are the

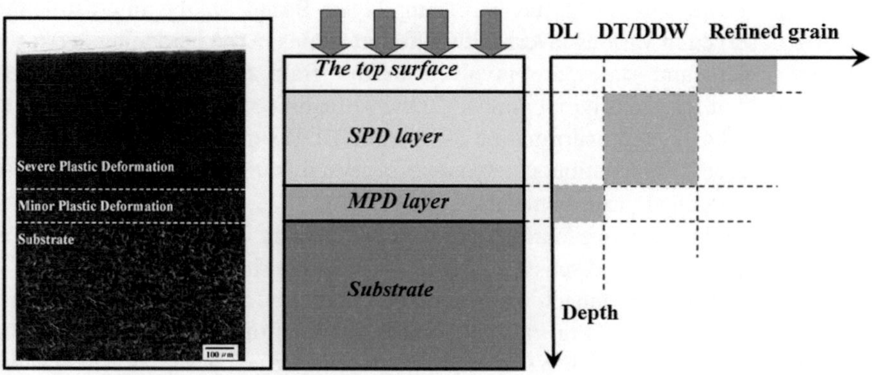


**Fig. 4.8** Schematic illustrations of microstructure characteristics along depth direction in the hardening layer subjected to three LSP impacts

magnitudes of the Burgers vectors of the perfect dislocation and the Shockley partial dislocation, respectively. The parameter $\alpha$ reflects the character of the dislocation and contains the scaling factor between the length of the dislocation source and the grain size.

The grain boundaries are taken as dislocation sources, as predicted by computer simulations for subgrains or refined grains. When the grain size becomes smaller than a critical value, $D_C$, determined by equating Eqs. 4.1 and 4.2

$$D_C = \frac{2\alpha\mu(b_N - b_P)b_P}{\gamma} \tag{4.3}$$

the generation of subgrain interfaces and stacking faults offers an alternative interpretation to dislocation pile-up at grain boundaries to explain the continuous grain-size strengthening, as suggested by Eq. 4.2, and the strain hardening of the metal materials.

The reaction between the laser shock wave and the sample will be generated near the sample surface, leading to the generation of the dislocation and the microstructural deformation near the surface, which can be explained by the fact that the compressive residual stresses is generated in the PD layer, and the magnitude of the compressive residual stress decreases away from the top surface.

### *4.5.2 Schematic Illustration of Grain Refinement Induced by Multiple LSP Impacts*

It is well known that the strain and strain rate decrease with the increasing depth from the top surface by LSP, various dislocation activities are normally motivated, including sliding, accumulation, interaction, tangling, and spatial rearrangement in order to accommodate plastic strains in Al alloys. In the Al alloy sample by multiple LSP impacts, dislocation activities lead to formation of DTs and DDWs in original coarse grains of the hardening layer. Based on the microstructure features observed in various layers with different strains in the hardening layer, the following elemental states are involved in the grain refinement process: (1) development of DLs in original grains; (2) the formation of DTs and DDWs due to the pile up of DLs; (3) transformation of DTs and DDWs into subgrain boundaries; (4) evolution of the continuous dynamic recrystallization (DRX) in subgrain boundaries to refined grain boundaries.

The grain refinement mechanism can be schematically illustrated in Fig. 4.8, in which each state will be discussed in terms of the experimental observations.

During the first LSP impact, dislocation activities lead to the formation of DLs in original grains [state (I) in Fig. 4.9], and the pile up of DLs contributes to the formation of the DTs and DDWs in original grains [state (II) in Fig. 4.9], as can be clearly seen in Fig. 4.7f and h. With further increasing strains, development of these dislocation configurations gradually results in subdivision of original grains

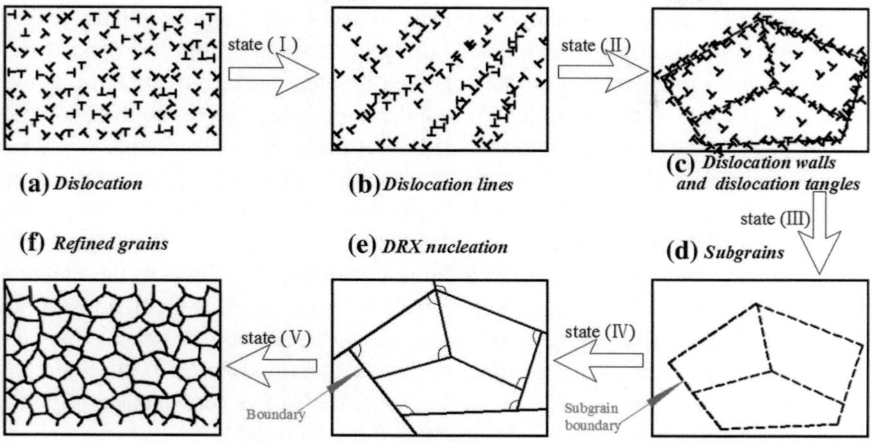


**Fig. 4.9** Schematic illustration showing microstructural evolution process of LY2 Al alloy induced by multiple LSP impacts

by forming individual dislocation cells primarily separated by DTs and DDWs [state (III) in Fig. 4.9], as can be clearly seen in Fig. 4.7f. At a certain strain level, for minimizing the total system energy, dislocation annihilation and rearrangement occur in DTs and DDWs, which will transform into subgrain boundaries separating individual cells [state (IV) in Fig. 4.9], as can be clearly seen from Figs. 4.9d and 4.7e. As concerned with the very high stacking fault energy of Al alloy and dislocation moving, the continuous dynamic recrystallization (DRX) may take place [state (V) in Fig. 4.9] in this process, that is to say, the DRX processes result in a progressive accumulation of boundary misorientation, and finally lead to a gradual transition of boundaries character until the formation of high angle grain boundaries[26]. As the top surface layer is deformed at a very high strain rate, the orientations of the grains with respect to their neighboring grains become completely random, and highly disoriented grain boundaries form, as can be clearly seen in Fig. 4.9e. As can be seen from Fig. 4.7f, subgrains with a width of 100–200 nm are formed, which eventually transform into equiaxed refined grains by development of subgrain boundaries.

During multiple LSP impacts, the second LSP impact may lead to the change of slip systems along depth direction even inside the same grain, so the grains can be subdivided efficiently by the DTs and DDWs. After multiple LSP impacts, the strain and strain rate further increase, and DTs and DDWs could form inside the inner of the refined grains, which indicates that the refined grains could be subdivided following the similar mechanism (i.e. the path Fig. 4.9a–f), as can be seen in Fig. 4.7g. With the increasing strain, the subdivision process takes place on a finer scale. When dislocation multiplication rate is balanced by the annihilation rate, the increase of strains could not reduce the grain size any longer, and a stabilized grain size is resulted. It is well known that at a given level of strain, dislocation density increases with an increasing strain rate. At a certain strain rate, an increase of strain leads to a higher density of dislocations and eventually to finer grains. But after multiple LSP impacts, the grain size could not be reduced infinitely, and the similar results can be achieved by other surface treatment methods [7, 8, 10]. The reason for obtaining the stabilized grain size might be that new DTs and DDWs could not be generated inside the refined grains. From Sect. 4.3, it can be seen that the residual stress distribution in depth direction can be seen during multiple LSP impacts. From Sect. 4.3, it can be concluded that the increasing rates of the depth with the SPD layer and the PD layer decrease as the impact time increases, and when the impact time is from four to five, the depth of the PD layer is almost not changed. This result agrees well with the previous studied result [16, 18]. These phenomenon can be attributed to the fact that after the balance between dislocation multiplication and annihilation at the certain strain rate, the grain size keep a stabilized value and the plastic deformation in the near-surface region is almost saturated after multiple LSP impacts (i.e., $\geq$ four).

During multiple LSP impacts, the strain rate in the drops remarkably with an increase of depth from the surface [27], and there are different dislocation structures in different depths, so the strain and strain rate play an important role in the grain refinement process and the final stabilized grain size upon plastic deformation. Table 4.3 shows the comparisons of strain, strain rate and the final grain size among severe plastic deformation techniques. It can be seen that there are clear difference in the average grain size by high pressure torsion [28], cold rolling [8], ball milling [9] and surface mechanical attrition treatment (SMAT) [10, 27, 29]. The minimum grain size by cold rolling and high pressure torsion is in the sub-micrometer range, while that by ball milling and SMAT is about 10 nm [27, 29]. This phenomenon can be attributed to the fact that the strain rate is about $10^3$ $s^{-1}$ at the top surface of the treated sample during ball milling and SMAT, but that is 10–102 s-1 in the top surface of the treated sample during high pressure torsion and cold rolling [8, 28].

It is worth noting that although the strain rate with about $10^7$ $s^{-1}$ in the top surface during multiple LSP impacts, which is far higher than that in the ball milling and SMAT, the refined grain size is the sub-micrometer range (100–200 nm). More systematic investigations are needed to demonstrate this phenomenon.

**Table 4.3** Comparisons of strain, strain rate and the final grain size among severe plastic deformation techniques

| Fabrication method | Strain rate ($s^{-1}$) | The minimum grain size (nm) |
|---|---|---|
| Cold rolling | $10 \sim 100$ | 100–200 |
| High pressure torsion | $10 \sim 100$ | 100–200 |
| Ball milling | $\sim 10^3$ | $\sim 10$ |
| SMAT | $\sim 10^3$ | $\sim 10$ |
| LSP | $\sim 10^7$ | 100–200 |

## 4.6 Summary

This chapter describes the effects of multiple LSP impacts on the residual stress and plastic deformation of the hardening layer of LY2 Al alloy, and presents the refined structure at the top surface of LY2 Al alloy subjected to multiple LSP impacts. Some important conclusions can be made as follows:

1. With increasing impact time during multiple LSP impacts, the surface compressive residual stress increases obviously. However, the increasing rate of residual stresses decreases as the impact time increases, and when the residual stresses at the locations far away from the surface (i.e., $\geq$ a few mm) are almost not influenced by the impact time.
2. The increasing rate of the affected depth in plastic deformation decreases with the increment of impact time, and when the impact time is from four to five, there are little changes in the depth of plastic deformation.
3. The grain in the hardening layer subjected to multiple LSP impacts is clearly refined, and the grain size is about 100–200 nm at the top surface.
4. Based on the microstructure observations, a grain refinement mechanism induced by plastic deformation during multiple LSP treatment in LY2 Al alloy is proposed. It involves the development of DLs in original grains, the pile up of DLs contributes to the formation of DTs and DDWs, the transformation of DTs and DDWs into subgrain boundaries, and the evolution of the continuous dynamic recrystallization (DRX) in subgrain boundaries to refined grain boundaries.
5. Experimental evidences and analysis of the grain refinement mechanism indicate that high strains with a high strain rate are necessary for the formation of refined grains during plastic deformation of LY2 Al alloy.

## References


1. Meyers, M. A., Gregori, F., Kad, B. K., Schneider, M. S., Kalantar, D. H., Remington, B. A., et al. (2003). Laser-induced shock compression of monocrystalline copper: characterization and analysis. *Acta Materialia, 51*(5), 1211–1228.
2. Zhang, H., & Yu, C. Y. (1998). Laser shock processing of 2024–T62 aluminum alloy. *Materials Science and Engineering A, 257*, 322–327.
3. Montross, C. S., Ye, L., Wei, T., Clark, G., & Mai, Y. W. (2002). Laser shock processing and its effects on microstructure and properties of metal alloys: a review. *International Journal of Fatigue, 24*, 1021–1036.
4. Zhang, Y. K., Hu, C. L., Cai, L., Yang, J. C., & Zhang, X. R. (2001). Mechanism of improvement on fatigue life of metal by laser-excited shock waves. *Applied Physics A, 72*(2), 113–116.
5. Yilbas, B. S., Shuja, S. Z., Arif, A., & Gondal, M. A. (2003). Laser-shock processing of steel. *Journal of Materials Processing Technology, 135*(1), 6–17.

6. Srinivasan, S., Garcia, D. B., Gean, M. C., Murthy, H., & Farris, T. N. (2009). Fretting fatigue of laser shock peened Ti–6Al–4 V. *Tribology International, 42*(9), 1324–1329.
7. Lindemann, J., Buque, C., & Appel, F. (2006). Effect of shot peening on fatigue performance of a lamellar titanium aluminide alloy. *Acta Materialia, 54*(4), 1155–1164.
8. Bhattacharjee, P. P., Ray, R. K., & Tsuji, N. (2009). Cold rolling and recrystallization textures of a Ni–5 at.% W alloy. *Acta Materialia, 57*(7), 2166–2179.
9. Venugopal, T., Rao, K. P., & Murty, B. S. (2007). Mechanical and electrical properties of Cu–Ta nanocomposites prepared by high-energy ball milling. *Acta Materialia, 55*(13), 4439–4445.
10. Lin, Y. M., Lu, J., Wang, L. P., Xu, T., & Xue, Q. J. (2006). Surface nanocrystallization by surface mechanical attrition treatment and its effect on structure and properties of plasma nitrided AISI 321 stainless steel. *Acta Materialia, 54*(20), 5599–5605.
11. Yang, C. H., Hodgson, P. D., & Liu, Q. C. (2008). Geometrical effects on residual stresses in 7050–T7451 aluminum alloy rods subject to laser shock peening. *Journal of Materials Processing Technology, 201*, 303–309.
12. Luong, H., & Hill, M. R. (2008). The effects of laser peening on high-cycle fatigue in 7085–T7651 aluminum alloy. *Materials Science and Engineering A, 477*, 208–216.
13. Gomez-Rosas, G., Rubio-Gonzalez, C., Ocaña, J. L., Molpeceres, C., Porro, J. A., Chi-Moreno, W., et al. (2005). High level compressive residual stresses produced in aluminum alloys by laser shock processing. *Applied Surface Science, 252*(4), 883–887.
14. Rubio-González, C., Ocaña, J. L., Gomez-Rosas, G., Molpeceres, C., Paredes, M., Banderas, A., et al. (2004). Effect of laser shock processing on fatigue crack growth and fracture toughness of 6061–T6 aluminum alloy. *Materials Science and Engineering A, 386*, 291–295.
15. Rubio-González, C., Gomez-Rosas, G., Ocaña, J. L., Molpeceres, C., Banderas, A., Porro, J., et al. (2006). Effect of an absorbent overlay on the residual stress field induced by laser shock processing on aluminum samples. *Applied Surface Science, 252*(18), 6201–6205.
16. Peyre, P., Fabbro, R., Merrien, P., & Lieurade, H. P. (1996). Laser shock processing of aluminum alloys. Application to high cycle fatigue behaviour. *Materials Science and Engineering A, 210*, 102–113.
17. Masse, J. E., & Barreau, G. (1995). Surface modification by laser induced shock waves. *Surface Engineering, 11*, 131–142.
18. Ding, K., & Ye, L. (2003). Three-dimensional dynamic finite element analysis of multiple laser shock peening processes. *Surface Engineering, 19*(5), 351–358.
19. Hu, Y. X., & Yao, Z. Q. (2008). Overlapping rate effect on laser shock processing of 1045 steel by small spots with Nd:YAG pulsed laser. *Surface and Coatings Technology, 202*, 1517–1525.
20. Schino, A. D., & Kenny, J. M. (2003). Grain size dependence of the fatigue behaviour of a ultrafine-grained AISI 304 stainless steel. *Materials Letters, 57*(21), 3182–3185.
21. Belyakov, A., Tsuzaki, K., Miura, H., & Sakai, T. (2003). Effect of initial microstructures on grain refinement in a stainless steel by large strain deformation. *Acta Materialia, 51*(3), 847–861.
22. Tan, Y., Wu, G., Yang, J. M., & Pan, T. (2004). Laser shock peening on fatigue crack growth behavior of aluminum alloy. *Fatigue and Fracture of Engineering Materials and Structures, 27*(8), 649–656.
23. Yilbas, B. S., & Arif, A. F. M. (2007). Laser shock processing of aluminum: Model and experimental study. *Journal of Physics D, Applied Physics, 40*, 6740–6747.
24. Chen, M., Ma, E., Hemke, K. J., Sheng, H. W., Wang, Y. M., & Cheng, X. M. (2003). Deformation twinning in nanocrystalline aluminum. *Science, 300*, 1275–1277.
25. Yamakov, V., Wolf, D., Phillpot, S. R., Mukherjee, A. K., & Gleiter, H. (2002). Dislocation processes in the deformation of nanocrystalline aluminium by molecular-dynamics simulation. *Nature Materials, 1*, 45–49.
26. Wen, M., Liu, G., Gu, J. F., Guan, W. M., & Lu, J. (2009). Dislocation evolution in titanium during surface severe plastic deformation. *Applied Surface Science, 255*(12), 6097–6102.

27. Tao, N. R., Wang, Z. B., Tong, W. P., Sui, M. L., Lu, J., & Lu, K. (2002). An investigation of surface nanocrystallization mechanism in Fe induced by surface mechanical attrition treatment. *Acta Materialia, 50*(18), 4603–4616.
28. Wang, Y. B., Louie, M., Cao, Y., Liao, X. Z., Li, H. J., Ringer, S. P., et al. (2010). High-pressure torsion induced microstructural evolution in a hexagonal close-packed Zr alloy. *Scripta Materials, 62*(4), 214–217.
29. Wu, X., Tao, N., Hong, Y., Liu, G., Xu, B., Lu, J., et al. (2005). Strain-induced grain refinement of cobalt during surface mechanical attrition treatment. *Acta Materialia, 53*(3), 681–691.

# Chapter 5
# Visual Inspection and Control Methods of Laser Sock Processing Effectiveness

**Abstract** This chapter introduces a novel visual inspection and control method to evaluate the surface qualities of Al alloy by means of three parameters, i.e., surface roughness, surface residual stress, and hollow depth, and also presents the effects of the surface qualities of LSP zones upon the fatigue life of samples and the optimum conditions of LSP.

## 5.1 Introduction

The possibility of using a pulsed, high-power laser beam to produce recoil pressures from vaporization of metal surfaces was presented by Askaryon et al. [1] in 1963. This effect on unconfined surfaces was verified by White [2]. The irradiated target was placed in a vacuum, the high-temperature plasma freely expanded, and shock waves induced at the target surface [3, 4]. Then Anderholm et al. [5] showed that pressures in the range of gigapascals (GPa) could be obtained at confined surfaces, i.e., surfaces covered by transparent overlay.

These high-amplitude shock waves can be used to improve the mechanical properties of metallic materials and alloys, especially, fatigue performance and fracture resistances; namely laser shock processing (LSP). In the early 1980s, Clauer et al. [6] studied LSP for improving fatigue resistances of aircraft aluminum alloys, such as 2024-T3, etc., with a large Nd: glass laser system with six amplifier stages. The results showed that the fatigue lives were increased considerably. In 1992, Vaccri [7] reported that the LSP could improve effectively the mechanical properties of many materials, for example, carbon and alloy steel, including powder-metals, stainless steels, malleable and nodular irons, aluminum alloy, titanium and titanium alloys, and nickel-base superalloy. The fatigue life of the tensile-fatigue specimen of 2024-T3 aluminum alloy was increased by four-fold after LSP [7]. In recent years, LSP applications have been unceasingly widely studied in France and in other countries [8–12]. But, if the selected laser

Y. Zhang et al., *Laser Shock Processing of FCC Metals*,
Springer Series in Materials Science 179, DOI: 10.1007/978-3-642-35674-2_5,

parameters and coating thicknesses are not suitable, the harmful heat damage will occur on the metal surfaces and their fatigue life will be decreased greatly.

In this chapter, the surface qualities of Al alloy are evaluated quantitatively by means of three parameters, i.e., surface roughness, surface residual stress, and hollow depth. Finally, good results which verify the validity of visual inspection and control methods are obtained. In addition, the effect of the surface qualities of LSP zones upon the fatigue life of samples and the optimum conditions of LSP are presented by using a moderate sized laser system.

## 5.2 Experiment Details

### 5.2.1 Selection of Laser Parameters

A Q-switched, Nd: glass laser with a wavelength of 1,064 μm and a pulse duration of 30 ns is used in the experiment. The laser system has three amplifier stages: a pre-set amplifier, a triple amplifier and a post-set amplifier. A storage oscilloscope 7,834 Tek with a resolution of time of 1 ns is used for monitoring the waveform of the laser at regular intervals. Eight percent of the laser beam energy is split using a splitter and transferred directly to a PT-lG energy meter for controlling the energy on-line.

The laser is capable of emitting up to 50 J of laser energy in a pulse that is approximately triangular in shape with a full width at half maximum of 30 ns. The laser beam is focused into a desired spot size on the sample surfaces with a 600 mm focal length double convex lens. The laser parameters used for LSP are as follows: pulse duration 30 ns, energy 8.1–34 J, spot size 0.7 mm, power density 0.7–2.9 GW $cm^{-2}$.

### 5.2.2 Samples and Treatments

The sample material is aluminum alloy 2024-T62 and its chemical composition is shown in Table 5.1, where $\delta_b$ and $\delta_{0.2}$ are the tensile strength and yield strength of 2024-T62, respectively. The fatigue samples are typical fasten-hole structure in the shape of a "dog bone", as shown in Fig. 5.1. The black material is coated at the surface of the sample, and a transparent K 9 glass with a diameter of 20 mm and a height of 4.5 mm or overlay is pressed against coating. For comparison, one side of the hole is shocked successively and the other is not shocked. After LSP, based on

**Table 5.1** The chemical compositions and mechanical properties of 2024-T62

| Composition | Cu | Mg | Mn | Si | Fe | Ti | Zr + Ti | Cr | In | Al |
|---|---|---|---|---|---|---|---|---|---|---|
| Percent (wt %) | 3.8–4.9 | 1.2–1.8 | 1.2–1.8 | 0.5 | 0.5 | 0.15 | <0.2 | 0.1 | 0.25 | Other |

*Mechanical properties* $\delta_b$, 425 MPa; $\delta_{0.2}$, 340 MPa; elengation ratio, 5 %; elastic modulus, 68.9 GPa; density, 2,770 kg/$m^3$ . Heat treatment, T62; solution heat treatment, artificial ageing

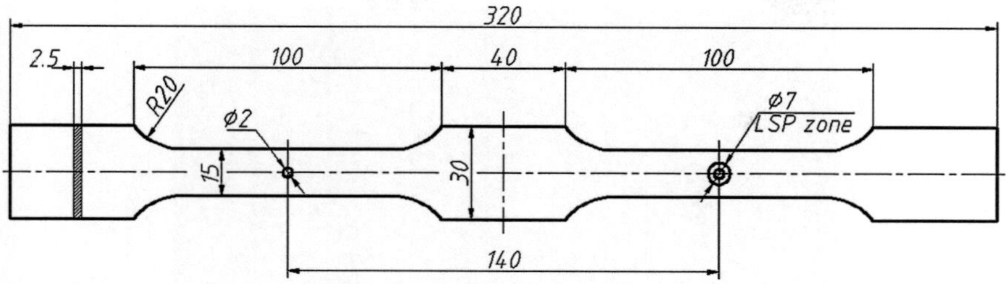


**Fig. 5.1** Profile and dimensions (mm) of the sample used in fatigue testing

analytic results of these samples, the surface qualities of the LSP zones can be divided into four grades, i.e., A. B, C and D. For comparison between sample and control gauge surface roughness, the surface roughness of the LSP zone is inspected visually. Then, the surface residual stresses are measured using X-ray diffraction.

Because the sample surfaces are damaged heavily in grades C and D, the surface residual stresses are not measured for grade C and D samples. The depth of the surface hollow is measured using a contourogram for grade A. Two fasten-holes, including an un-shocked hole with a diameter of 2 mm, are bored (based on a center hole of $\Phi$ 1 mm) using a Hauser $OP_2$ precision jig boring machine. The roughness should be less than 0.1 μm at the inside surface of hole. All the LSP samples are tested to failure at a maximum stress loading of 4.2 kN and R = 0.1 under constant-amplitude load control with a precision of +0.5 %, and the test frequency is 13 Hz, i.e., low frequency fatigue failure.

## 5.3 Surface Quality and Fatigue Life

### *5.3.1 Grade A Surface Quality and Fatigue Life*

The surface features of the grade A surface quality show that the surface roughness of the LSP zone is less than or equal to the surface roughness of the un-processed area, a very dense bright shallow hollow is formed and a shot spot is well-defined, as shown in Fig. 5.2. Among these LSP samples, as shown in Table 5.2, ten samples achieve grade A surface quality. The fatigue lives of the grade A samples are shown in Table 5.3.

### *5.3.2 Grade B Surface Quality and Fatigue Life*

The surface features of the grade B surface quality show that surface materials are vaporized uniformly or there are a few craterlets produced by vaporization at the LSP zone, as shown in Fig. 5.3. All five samples have grade B surface quality, as shown in Table 5.4. The fatigue lives are shown in Table 5.5.

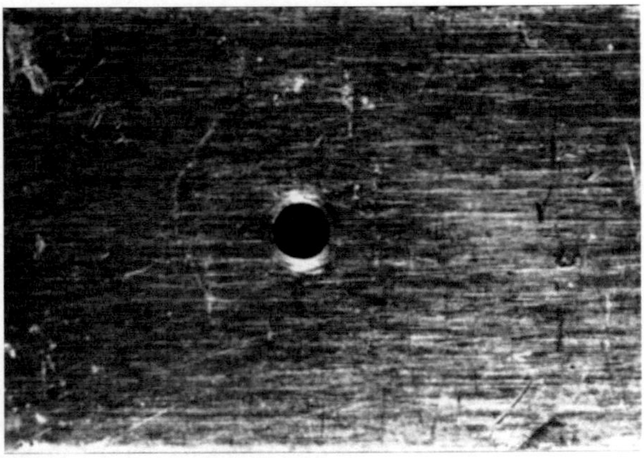

**Fig. 5.2** Photograph of grade A surface quality (3.3 x)

**Table 5.2** Surface features of samples with grade A surface quality

| Sample no. | Surface roughness (μm) | | Surface residual stress (MPa) | | Hollow depth (μm) |
|---|---|---|---|---|---|
| | Un-shocked zone | LSP zone | Un-shocked zone | LSP zone | |
| I-1 | 3.2 | 3.2,1.6 | 0 | −60,−60 | 10,6 |
| I-2 | 3.2 | 0.8,0.8 | 0 | −70,−72 | 16,9 |
| I-3 | 3.2 | 3.2,3.2 | 0 | −60,−40 | 8,7 |
| I-4 | 3.2 | 0.8,0.8 | 0 | −60,−70 | 10,14 |
| I-5 | 3.2 | 0.8,0.8 | 0 | −64,−62 | 16,16 |
| I-6 | 3.2 | 0.8,0.8 | 0 | −68,−70 | 16,17 |
| I-7 | 3.2 | 3.2,1.6 | 0 | −40,−50 | 8,9 |
| I-8 | 3.2 | 0.8,0.8 | 0 | −66,−76 | 18,18 |
| I-9 | 3.2 | 3.2,1.6 | 0 | −42,−52 | 10,11 |
| I-10 | 3.2 | 1.6,0.8 | 0 | −72,−70 | 17,15 |

**Table 5.3** The results of fatigue testing for grade A surface quality

| Sample no. | Cycles of: | | Increase (%) $(N_{bi} - N_{ai})/N_{ai}$ |
|---|---|---|---|
| | Unshocked samples ($N_{ai}$) | LSP samples ($N_{bi}$) | |
| I-1 | 96,580 | 356,930 | 269.4 |
| I-2 | 77,710 | 755,870 | 872.7 |
| I-3 | 73,850 | 297,750 | 303.2 |
| I-4 | 88,630 | 719,640 | 711.9 |
| I-5 | 93,580 | 723,080 | 662.7 |
| I-6 | 73,700 | 567,610 | 670.2 |
| I-7 | 98,000 | 414,270 | 322.7 |
| I-8 | 95,340 | 823,460 | 763.7 |
| I-9 | 81,480 | 359,360 | 341.0 |
| I-10 | 107,320 | 799,840 | 645.3 |
| Average | | | 556.3 |

**Fig. 5.3** Photograph of grade B surface quality (3.3 x)

**Table 5.4** Surface features of samples with grade B surface quality

| Sample no. | Surface roughness (μm) | | Surface residual stress (MPa) | |
|---|---|---|---|---|
| | Unshocked zone | LSP zone | Unshocked zone | LSP zone |
| II-1 | 3.2 | 6.3,6.3 | 0 | −20,−10 |
| II-2 | 3.2 | 6.3,6.3 | 0 | −14,−12 |
| II-3 | 3.2 | 12.5,12.5 | 0 | −14,−10 |
| II-4 | 3.2 | 12.5,12.5 | 0 | −10,−16 |
| II-5 | 3.2 | 12.5,6.3 | 0 | −12,−14 |
| II-6 | 3.2 | 12.5,12.5 | 0 | −16,−16 |

**Table 5.5** The results of fatigue testing for grade B surface quality

| Sample no. | Cycles of: | | Increase (%) $(N_{bi} - N_{ai})/N_{ai}$ |
|---|---|---|---|
| | Unshocked samples ($N_{ai}$) | LSP samples ($N_{bi}$) | |
| II-1 | 96,580 | 356,930 | 269.4 |
| II-2 | 77,710 | 755,870 | 872.7 |
| II-3 | 73,850 | 297,750 | 303.2 |
| II-4 | 88,630 | 719,640 | 711.9 |
| II-5 | 93,580 | 723,080 | 662.7 |
| II-6 | 73,700 | 567,610 | 670.2 |
| Average | | | 556.3 |

### *5.3.3 Grade C Surface Quality and Fatigue Life*

The surface features of grade C surface quality show there are dispersed, honeycomb-like, uneven, big vaporized spots and ray-like ditches which center around the holes and radiate outside of the LSP zone, as shown in Fig. 5.4. All four samples have grade C surface quality, and the changes of surface roughness are shown in Table 5.6. The fatigue lives are shown in Table 5.7.

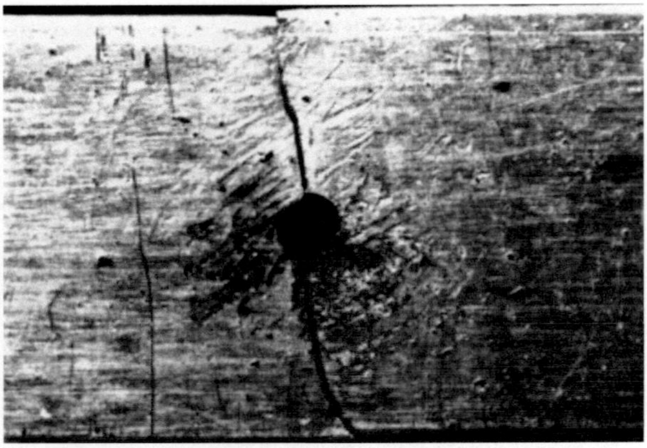

**Fig. 5.4** Photograph of grade C surface quality (3.3 x)

**Table 5.6** Surface features of samples with grade C surface quality

| Sample no. | Surface roughness (μm) | |
|---|---|---|
| | Unshocked zone | LSP zone |
| III-1 | 3.2 | 25, 25 |
| III-2 | 3.2 | 25, 12.5 |
| III-3 | 3.2 | 25, 25 |
| III-4 | 3.2 | 12.5, 25 |

**Table 5.7** The results of fatigue testing for grade C surface quality

| Sample no. | Cycles of: | | Increase (%) $(N_{bi} - N_{ai})/N_{ai}$ |
|---|---|---|---|
| | Unshocked samples ($N_{ai}$) | LSP samples ($N_{bi}$) | |
| III-1 | 70,700 | 67,700 | −4.2 |
| III-2 | 98,190 | 105,070 | 7 |
| III-3 | 83,610 | 77,930 | −6.8 |
| III-4 | 89,570 | 96,110 | 7.3 |
| Average | | | 3.3 |

### *5.3.4 Grade D Surface Quality and Fatigue Life*

The surface features of grade D surface quality show that the surface metals are vaporized intensely and melt, the grains and surface roughness are bigger, and the LSP zone becomes a big ablated concave with many ray-like ditches, as shown in Fig. 5.5. All three samples have grade D surface quality, and the changes of the surface roughness are shown in Table 5.8. The fatigue lives are shown in Table 5.9.

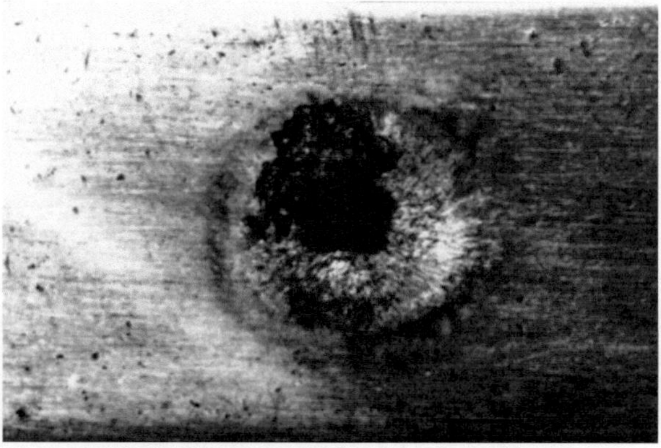

**Fig. 5.5** Photograph of grade D surface quality (3.3 x)

**Table 5.8** Surface features of samples with grade C surface quality

| Sample no. | Surface roughness (μm) | |
|---|---|---|
| | Unshocked zone | LSP zone |
| IV-1 | 3.2 | 50,50 |
| IV-2 | 3.2 | 50,50 |
| IV-3 | 3.2 | 50,50 |

**Table 5.9** The results of fatigue testing for grade C surface quality

| Sample no. | Cycles of: | | Increase (%) $(N_{bi} - N_{ai})/N_{ai}$ |
|---|---|---|---|
| | Unshocked samples ($N_{ai}$) | LSP samples ($N_{bi}$) | |
| IV-1 | 91,210 | 67,010 | −26.5 |
| IV-2 | 88,910 | 47,170 | −47.0 |
| IV-3 | 82,470 | 32,640 | −70.0 |
| Average | | | −44.6 |

## 5.4 Analysis and Discussion

### *5.4.1 Fatigue Life*

For grade A surface quality, the fatigue lives of ten samples are increased by 556.3 % on average. With 95 % confidence, the mean fatigue life of LSP samples 95 % confidence, the mean fatigue life of LSP samples is 4.8–8.1 times larger than un-shocked samples. Experimental results show less scatter and high reliability.

For grade B surface quality, the fatigue lives of six samples are increased by 68.4 % on average. With 95 % confidence, the mean fatigue life of LSP samples is 1.14–1.60 times larger than un-shocked samples.

For grade C surface quality, the fatigue lives of four samples are increased by 3.3 %. For grade D surface quality, the fatigue lives of three samples are decreased by 44.6 % on average.

### 5.4.2 Surface Qualities

As the metallic materials are impinged under the high pressures of shock waves, the plastic deformation is induced to form an LSP strengthening zone. The macroscopic features of sample surfaces display high-level residual compressive stress and high hardness. The inside microscopic microstructures are dense dislocations and retied grains. These factors combine to improve greatly the fatigue lives of materials, which can indicated by means of the surface roughness and hollow shape at the LSP zone. The smaller the surface roughness, the better the LSP effect is. The depth and shape of the hollow gives some indication about the extent of the dislocation density, surface hardness and residual stress affected area. In general, the hollow is a small plastic deformation of only 6–18 μm in depth, and can be observed visually by reflection of a light beam. The grade A surface quality mainly reflects the LSP contribution to fatigue life. The grade B surface quality rather reflects LSP contribution to fatigue life, but a part of the fatigue life gain is lost because of surface quality degradation. The grade C surface quality cannot reflect LSP contribution to fatigue life, and the fatigue life gain is about equal to fatigue life loss caused by surface quality degradation.

The grade D surface quality cannot reflect LSP contribution to fatigue life, the fatigue life gain cannot compensate enough fatigue life loss caused by surface quality degradation, and thus the fatigue life increment of LSP sample is negative. The harmful heat damage produced by LSP is completely avoided in grade A surface quality, but the heat damage does exist to some extent in the others.

### 5.4.3 How to Obtain Grade A Surface Quality

If the peak pressure of shock waves induced by LSP is higher than the dynamic yield strength, the shock waves can cause plastic deformation on the surface layer, the laser parameters are selected according to this.

In order to avoid producing grades B, C and D surface qualities in practical operations: first, select a thicker coating and smaller laser power densities for the LSP, by means of theoretical estimation; then, adjust the laser power densities and thickness of coating, as well as other process details to be optimized. Thus, grade A surface quality is obtained and controlled stably: namely, the shot spot is well-defined, the surface roughness of the LSP zone is less than or equal to the surface roughness of the un-shocked area, and a very bright shallow hollow is formed. Further experimental results show this procedure is very practical, which solves a difficult problem of LSP engineering applications, and ensures that the fatigue life of all samples is increased greatly after LSP.

## 5.5 Summary

The surface qualities of LSP zones are in close relationship with the fatigue lives of LSP samples. Only if the surface qualities of LSP zones achieve grade A, can the LSP greatly increase the fatigue lives of aluminum alloy 2024T-62. With 95 % confidence, the mean fatigue life of LSP samples with grade A surface qualities is 4.8–8.1 times larger than unprocessed ones. In practical operation, the grade A surface quality can be obtained by means of adjusting the laser parameters and the thickness of coating, as well as other process details. Additionally, in order to obtain the precise relationship between the surface quality produced by LSP and the fatigue life of metals, many in depth experiments are under way.

## References


1. Askaryon, G. A., & Morez, E. M. (1963). Pressure on evaporation of matter in a radiation beam. *JETP Letters, 16*, 1638–1644.
2. White, R. M. (1963). Generation of elastic waves by transient surface heating. *Journal of Applied Physics, 34*(12), 3559–3567.
3. Nuckolls, J., Wood, L., Thiessen, A., & Zimmerma, G. (1972). Laser compression of matter to super-high densities: Thermonuclear (CTR) applications. *Nature, 239*, 139–142.
4. Burckner, K. A., & Jorna, S. (1974). Laser-driven fusion. *Reviews of Modern Physics, 46*(2), 325–367.
5. Anderholm, N. C. (1970). Laser-generated stress waves. *Applied Physics Letters, 16*(3), 113–115.
6. Ford, S. C., Fairand, B. P., Clauer, A. H., & Galliher, R. D. (1980). Technical Report AFWAL-TR-80-3001 (Vol. 1, Aug 1980).
7. Vaccari, J. A. (1992). Laser shocking extends fatigue life. *American Machinist, 7*, 62–64.
8. Banas, H. E., Elsayed-Ah, H. E., Lawrences, F. V., & Rigsbee, J. M. (1990). Laser shock-induced mechanical and microstructural modification of welded maraging steel. *Journal of Applied Physics, 67*(5), 2380–2384.
9. Gerland, M., Hallouin, M., & Presles, H. N. (1992). Comparison of two new surface treatment processes, laser-induced shock waves and primary explosive: Application to fatigue behavior. *Materials Science and Engineering A, 156*(2), 175–182.
10. Grevey, D., Maffredy, L., & Vannes, A. B. (1992). Laser shock on a TRIP alloy: Mechanical and metallurgical consequences. *Journal of materials science, 27*(8), 2110–2116.
11. Dubrujeaud, B., Vardavoulias, M., & Jeandin, M. (1994). Dry sliding wear behaviour of a P/M ferrous alloy superficially densified by laser shock processing. *Surface and Coatings Technology, 67*(1–2), 125–132.
12. Masse, J. E., & Barreau, G. (1995). Laser generation of stress waves in metal. *Surface and Coatings Technology, 70*(2–3), 231–234.

# Chapter 6
# Mechanical Properties of AISI 304 SS and its Welded Joint Subjected to Laser Shock Processing

**Abstract** This chapter presents the effects of a single LSP impact on the nano-hardness, elastic modulus, residual stress, and phase transformation of AISI 304 SS. Further studies of different LSP paths on the mechanical properties and fracture morphology of the laser welded AISI 304 SS joint are presented and discussed.

## 6.1 Introduction

AISI 304 stainless steel (AISI 304 SS) is a widely used engineering material, which is suitable in marine regions due to its high oxidation resistance, but it cannot be hardened by heat treatments [1].

Due to the excellent mechanical and anti-corrosion properties of AISI 304 SS, and the ease with which it can be formed and welded, austenitic stainless steel has been widely used throughout the chemical, automotive and nuclear industries for reactor coolant piping, value bodies, and vessel internals [2]. In practical applications, AISI 304 SS is often required to be welded, and the poor mechanical properties of the stainless steel welding line limit its wide application [3]. For example, the tungsten inert gas (TIG) welding with the shallow penetration restricts its ability to weld thick structures in a single pass [4]. The conventional arc welding is often sensitive to form the coarse grains and intergranular Cr-rich carbides along the grain boundaries in heat-affected zone (HAZ), which deteriorates the mechanical properties of the joints [5, 6].

Laser welding possesses some advantages, including smaller deformation of workpiece and HAZ, deeper and narrower fusion zone (FZ), lower tensile residual stress and weld distortion, higher accuracy and degree of automation, over other welding processes [7, 8]. In order to achieve good welding quality, the combination of the laser welding parameters, such as the laser power, welding speed, focal length, shielding gas, position accuracy and so on, should be selected

Y. Zhang et al., *Laser Shock Processing of FCC Metals*,
Springer Series in Materials Science 179, DOI: 10.1007/978-3-642-35674-2_6,

according to the absorption and reflection of laser beam as well as the material thermo-physical properties [9, 10]. But the laser welded material may not attain the perfect outcome on occasion and the defects exist. For example, the porosity and crack are easy to be generated in the weld during laser welding. The welding joint may be softened and the strength decreased due to the uneven volatilization of the elements in the molten pool. The material may be deformed due to its thermal effect [11–14]. In addition, the cost of laser welding is expensive. With increasing concern of environmental safety and reduction of material consumption, it is necessary to employ green and safe technologies to optimize these inferior laser welded material joint. Most of all, it is important to improve its mechanical properties.

LSP is a recently developed technique which has been successfully applied in improving fatigue life and strength of metal material by generating the greater depth of compressive residual stress comparable to shot peening or cold rolling [15, 16]. Effects of LSP on the pure mechanical property or micro-structure feature of AISI 304 SS have attracted great attention of many researchers [17–21]. For instance, LSP without protective coating has been applied to water-immersed AISI 304 and AISI 316L SS. As a result, LSP can effectively retard crack initiation and growth in austenitic stainless steels due to the fact that the surface residual stress of both materials was converted from tensile stress to compressive residual stress of several hundred MPa after LSP [17]. The high temperature (up to 600 °C) fatigue behavior and residual stress stability of LSP-shocked (LSPed) AISI 304 SS were investigated, and the results demonstrated that LSP can enhance the life time [18]. The effects of a single LSP impact on residual stress relaxation and hardness of AISI 304 SS were investigated, and the results showed that LSP can improve the distributions of residual stress and micro-hardness [19]. Residual stress state of AISI 304 SS with and without LSP was also investigated, and deformation-induced martensitic transformation was found after LSP [20]. Micro-structural evolution and grain refinement in AISI 304 SS subjected to multiple LSP impacts were investigated, but it was not concerned with the corresponding mechanical properties [21]. These researches have highly focused on the improvement of micro-hardness and residual stress, fatigue, or micro-structure. Actually, the transformation in microstructure is always accompanied by a substantial increase in mechanical properties. Nevertheless, up to now, a clear scenery between microstructure and mechanical property of the plastic deformation layer for the LSPed austenitic stainless steel is still lacking. Therefore, a detailed understanding of micro-structural strengthening mechanism of LSP on mechanical property is crucial not only for practical application but also the nature of stainless steel. Little attention has been done on the elastic modulus and nano-hardness distribution of the samples with and without LSP under different maximum loads. In addition, although laser welding possesses the above-mentioned advantages, its thermal effect brings some disadvantages, such as surface tensile residual stress and porous microstructure. LSP is an effective method to eliminate tensile residual stress and refine the coarse grains in the surface layer of metallic materials and alloys.

With the above background in mind, the first purpose of this chapter is to investigate the effects of a single LSP impact on the nano-hardness, elastic modulus, residual stress, and phase transformation of AISI 304 SS. The elastic modulus and nano-hardness distribution of the samples with and without LSP under different maximum loads are investigated and discussed. Especially, the understanding of the corresponding relation between mechanical properties and micro-structural observations of AISI 304 SS subjected to a single LSP impact is investigated. The underlying enhancement mechanisms of LSP on nano-hardness, elastic modulus and residual stress of AISI 304 SS are also revealed by micro-structural observations. The second purpose of this chapter is to investigate the effects of different LSP paths on the mechanical properties and fracture morphology of the laser welded AISI 304 SS joint.

## 6.2 Measurement and Characteristics of Mechanical Properties and Micro-Structures

### *6.2.1 Experimental Material*

The chemical composition and mechanical properties of AISI 304 SS were shown in Tables 6.1 and 6.2, respectively. AISI 304 SS samples were cut into a rectangular shape with dimensions of 5 × 5 × 2 $mm^3$ (width × length × thickness) during a single LSP impact. AISI 304 SS plate with a thickness of 5 mm was used as the welding materials in this experiment.

Prior to LSP, the samples were polished by SiC paper with different grades of roughness (150-1500) in order to ensure uniform finishing, followed by cleaning in deionized water. Ultrasound in ethanol was used to degrease the sample surface.

**Table 6.1** The chemical composition of austenitic stainless steel AISI 304 (wt %)

| Composition | C | Mn | Cr | Mo | Ni | Cu | Si | Nb | Fe |
|---|---|---|---|---|---|---|---|---|---|
| Percent (wt %) | 0.06 | 1.54 | 18.47 | 0.30 | 8.3 | 0.37 | 0.48 | 0.027 | other |

**Table 6.2** Mechanical properties of AISI 304 SS

| Type | Value |
|---|---|
| Specific gravity, d, $g/cm^3$ | 7.93 |
| Tensile strength, $\sigma_b$, $kgf/mm^2$ | 520 |
| Yield strength, $\sigma$, $kgf/mm^2$ | 205 |
| Elongation, $\delta$, % | 40 |
| Vickers-hardness, HV | 200 |

**Table 6.3** Laser welding parameters

| Laser type | Laser power (kW) | Welding speed (mm/s) | Defocusing distance (mm) | Flow rate ($m^3$/h) |
|---|---|---|---|---|
| Fiber laser | 4 | 40 | −3 | 0.6 |

### *6.2.2 Laser Welding Parameters*

Laser welding was carried out using the YLS-4000 fiber laser (*IPG, USA*). On the side, the YW 50 laser welding head (*Precitec Germany*) was mounted on an ABB IRB 4400 robotic arm. In this laser welding test, the corresponding common welding parameters were chosen and shown in Table 6.3. The fiber laser power was 4 kW, the spot diameter of the laser beam was 0.27 mm with a focal length of 190 mm, and the defocusing distance $\triangle f$ was −3 mm. The ultra-high mixed gas integrating helium (He) and argon (Ar) (mixing ratio was 1:1) was selected as the shielding gas. The flow rate of adding gas was 0.6 $m^3$/h in the condition of coaxial adding gas during the laser welding. A common welding speed of 40 mm/s was conducted during the laser butt welding.

Figure 6.1 provides a schematic representation of the completed laser welded sample with the dimension of 80 × 100 × 5 mm. The tensile samples with 80 mm in length, 20 mm in width and 5 mm in height were cut from the acceptable laser welded plate. In such a way the tensile samples could contain portions of the HAZ and FZ in the center of the laser welded plate, as shown in Fig. 6.2a. The tensile samples were polished with 150-1500# SiC sand paper to avoid the effects of the surface roughness.

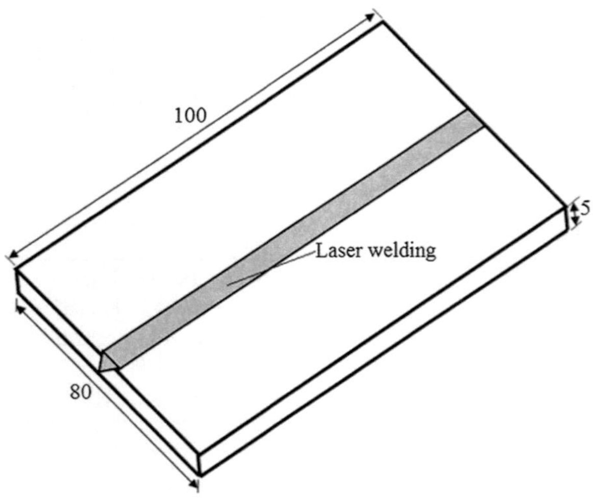


**Fig. 6.1** Schematic representation of sample used in laser welding (*unit* mm)

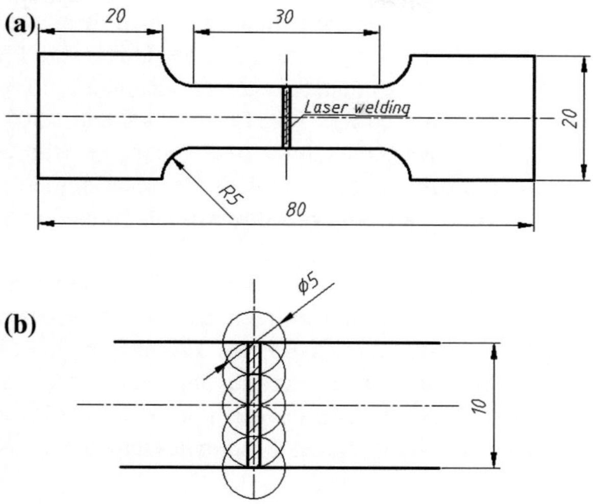


**Fig. 6.2** Dimensions of the tensile sample subjected to LSP (*unit* mm). **a** the laser welded tensile sample, and **b** the laser shocked paths of the welded sample

### 6.2.3 LSP Parameters

During a single LSP impact on AISI 304 SS and the LSP experiment of laser welded AISI 304 SS joint, LSP was carried out by a laser device at Laser Technology Institute in Jiangsu University. The shockwaves were induced by a Q-switched repetition-rate laser with a wavelength of 1,054 nm and a pulse width of around 20 ns. A water film with a thickness of 1–2 mm was used as a transparent confining layer and the professional 3 M aluminum tape (Made in USA) with a thickness of 100 μm was used as an absorbing layer to protect the sample surface from thermal effect. The laser energy was around 25 J, and the laser spot diameter was 5 mm.

The laser welded joint of the tensile samples was treated by single-sided LSP impacts and two-sided LSP impacts. The overlapping rate of the laser spot was 50 % and the treated area subjected to LSP on one side of the tensile sample can be seen in Fig. 6.2b. During the process of two-sided LSP, the high energy laser pulse was focused onto two sides of the welded tensile sample in turn.

### 6.2.4 Measurements of Nano-Hardness, Elastic Modulus and Residual Stress

The measurements of nano-hardness and elastic modulus for the samples with and without a single LSP impact were performed by nano-indentation technique on a *Nano Indenter ® XP, Hystron Corporation, USA* by using a *Berkovich* diamond indenter. The instrument can continuously measure the loading force and displacement and automatically calculate the projected area of contact according to the tip area function supplied. The load-displacement plot was then used to determine the hardness in terms of the Oliver and Pharr [22] method. The elastic modulus can be calculated by the corresponding formula. In the present work, the maximum load used was 500, 1,000, 1,500 and 1,500 μN, respectively. Therefore, the load-displacement curves of the samples with and without LSP at different maximum loads can be presented. The measurements were repeated five times for each condition, an average value was determined on the basis of the five measured data.

The surface residual stresses of the samples with and without LSP were determined using X-ray diffraction. The X-ray beam diameter was about 2 mm. The X-ray source was Cr-K$\alpha$ ray. The feed angle of the ladder scanning was 0.1° per second. The scanning starting angle and terminating angle were 145 and 153°, respectively. For the measurement of the residual stress along the depth direction, the electropolishing material removal method was used.

### 6.2.5 Measurements of Tensile Properties

Before the tensile test, the tensile samples after LSP were cleaned in deionized water and degreasing in ethanol by ultrasonic vibration. The gauge area of the tensile samples was carefully ground along the loading direction with sand papers up to a grit number of 600. There were 15 samples used in the tensile tests for single-sided LSP impacts and two-sided LSP impacts, respectively. The tensile test was performed on a MC electronic tensile testing machine at room temperature. The mechanical properties of the tensile samples subjected to single-sided LSP and two-sided LSP, such as time-loading curve, displacement-loading curve, etc., were obtained.

### 6.2.6 Micro-Structural Observations

A Rigaku D/max-2400 X-ray diffractometer (12 kW) with Cu/K$\alpha$ radiation was used to determine the phase constitution. The microstructures in the near-surface of the samples with and without LSP were observed by optical microscopy (OM) and transmission electron microscope (TEM). After LSP, the samples used for metallographic investigation were cut as the sections perpendicular to the sample surface, and then subjected to several successive steps of grinding and polishing. After that, the vertical sections of the samples were etched using a professional reagent that consists of 15 cc of HCl, 10 cc of $HNO_3$, 10 cc of acetic acid, and 2/3 drops glycerine, and then characterized by cross-sectional OM observations.

The micro-structures in the near-surface layer of the samples with and without a single LSP impact (the treated and untreated samples by a single LSP impact) were characterized by using a JEM-2100 TEM operated at a voltage of 200 kV. TEM foils in the near-surface layer were prepared by a combination of single and twin-jet electropolishing. All micrographs presented were bright-field images recorded under two-beam conditions.

The fracture morphology of the tensile samples subjected to single-sided LSP and two-sided LSP were examined by using a scanning electron microscopy (SEM) (JSM-6360LA, JEOL, Japan), respectively. The surface smoothness of the fracture morphology, the size, distribution and shape of the dimples were comparatively analyzed.

## 6.3 Nano-Hardness, Elastic Modulus and Residual Stress of AISI 304 SS

### *6.3.1 Effects of LSP on Nano-Hardness, Elastic Modulus and Residual Stress*

Figure 6.3a–d show the measured load-displacement curves of the samples with and without a single LSP impact at different maximum loads of 500, 1,000, 1,500, and 2,000 μN, respectively. It can be found that the contact depths of the indenter of the untreated samples are 61.52, 102.71, 138.78, and 146.44 nm, while those of the treated samples are 50.26, 81.72, 113.03, and 135.85 nm at different maximum loads of 500, 1,000, 1,500, and 2,000 μN, respectively. It can be clearly seen that the maximum contact depths of the indenter at peak load of the untreated samples is obviously higher than those of the treated samples at the same maximum load. Furthermore, the maximum contact depth increases with the increment of the maximum load in both cases.

From Fig. 6.3a–d, the values of nano-hardness and elastic modulus of the samples with and without a single LSP impact at different maximum loads are shown in details, respectively. It can be found that the values of nano-hardness for the untreated samples are 3.54, 2.92, 2.50, and 2.38 GPa, while those of the treated samples are 4.70, 3.57, 3.27, and 2.68 GPa, respectively. The values of elastic modulus of the untreated samples are 54.79, 41.89, 33.52, and 30.52 GPa, while those of the treated samples are 95.82, 72.19, 57.22, and 54.07 GPa, respectively.

Figure 6.4 exhibits the profiles of the surface nano-hardness and elastic modulus of the samples with and without a single LSP impact as a function of the maximum load with 500, 1,000, 1,500, and 2,000 μN. The following results can be clearly seen from Fig. 6.4. On the one hand, the values of nano-hardness and elastic modulus of the treated sample are obviously higher than those of the untreated samples at the same maximum load. On the other hand, the values of nano-hardness and elastic modulus of the samples with and without LSP decrease with the increment of the maximum load, and this phenomenon may be explained by the fact that the contact depth of the indenter increases with the increase of the maximum load, that is to say, the nano-mechanical property gradually decreases

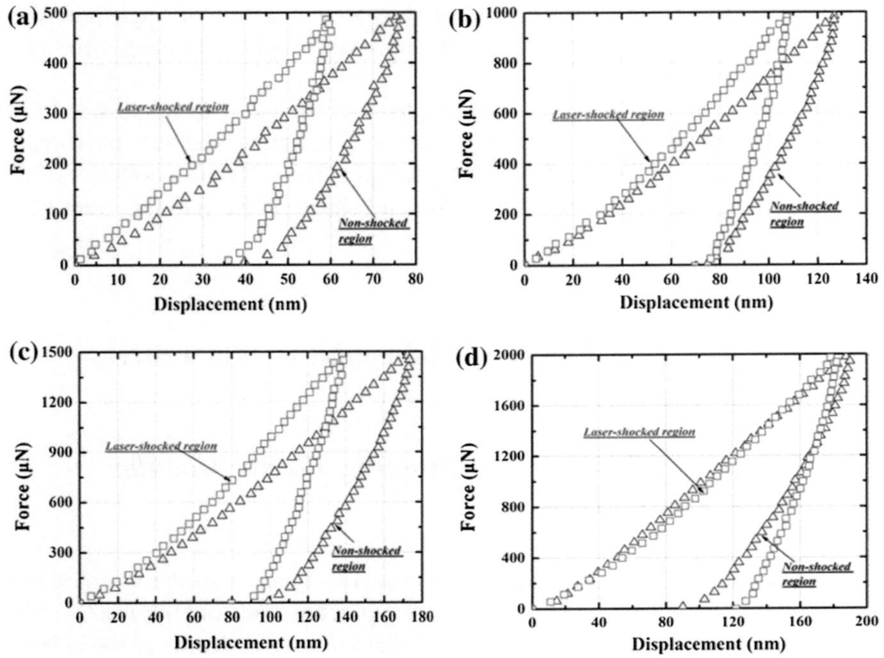


**Fig. 6.3** The measured load-displacement curves for the samples with and without a single LSP impact at different maximum loads of **a** 500, **b** 1,000, **c** 1,500, and **d** 2,000 μN, respectively

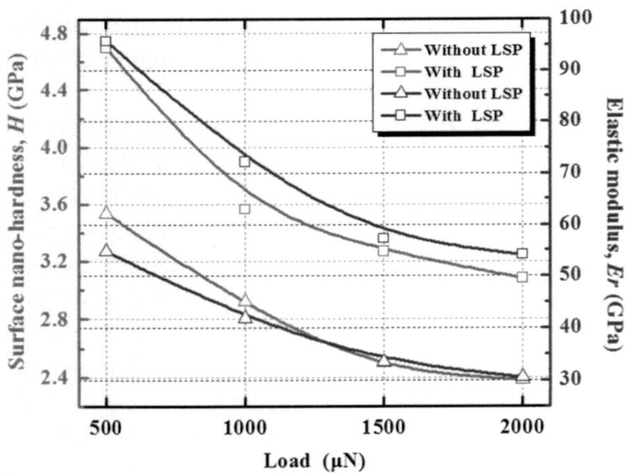


**Fig. 6.4** The profiles of the surface nano-hardness and elastic modulus of the samples with and without a single LSP impact as a function of the maximum load

with the increment of the distance to the treated surface. Note that with the increase of the maximum load, the decreasing rate of elastic modulus and nano-hardness for the treated sample is markedly higher than that for the untreated sample.

Although the elastic modulus is an intrinsic material property and fundamentally related to atomic bonding, the elastic modulus can be changed by some surface treatment technologies. The elastic modulus of AISI 304 SS sample by the high-energy shot peening and ultrasonic shot peening shows obviously a fluctuation along the depth direction [23]. It is well known that the increment of elastic modulus is favorable in enhancing the stiffness of the parts, which decides the stability when the components service in environment [24]. Hence, it can be concluded that LSP can improve the stiffness of AISI 304 SS sample. The surface hardness of metal material plays an important role in the resistance of foreign object damage (FOD). Generally speaking, the higher the surface hardness is, the stronger the resistance of FOD is [25, 26].

From the above experimental data, it can be seen that the values of nano-hardness and elastic modulus for the treated sample are greater than those in the untreated sample. Therefore, LSP can improve the nano-hardness and elastic modulus in the near-surface layer of the sample, which was favorable in improving FOD resistance of AISI 304 SS components.

Residual stress profiles of the samples with and without a single LSP impact in depth direction are presented in Fig. 6.5. It can be seen that the untreated regions are approximately in the zero-stress state, indicating that the effect of initial stress on the shock waves may be ignored [27]. High-level compressive residual stress is generated by a single LSP impact in the near-surface layer and remains in the compressive residual stress state up to a depth of approximately 900 μm. After LSP, the maximum residual stress of −305 MPa locates at the treated surface, and the value of the compressive residual stress decreases gradually with the increase the distance to the treated surface. Note that the increasing rate of surface compressive residual stress when the distance to the treated surface ranges from 0 to 200 μm is obviously higher than that when the distance to the treated surface ranges from 200 to 900 μm.

After LSP, the compressive residual stress exists in the near-surface layer, and the maximum compressive residual stress locates at the treated surface. It should

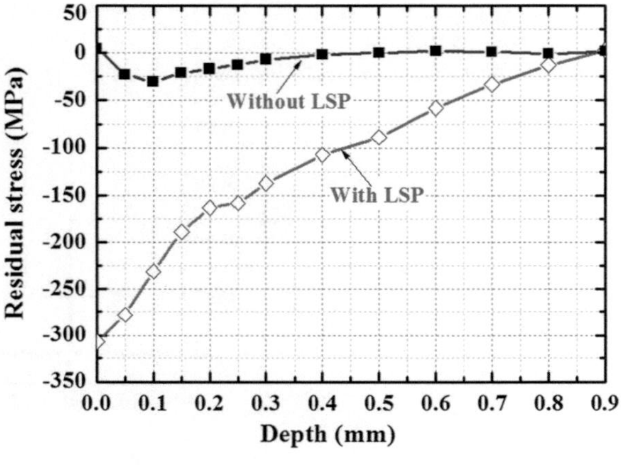


**Fig. 6.5** Residual stress profiles of the samples with and without a single LSP impact in depth direction

be noted that after shot peening there is a slight increase of compressive residual stress at the beginning stage, and the maximum compressive residual stress locates at the depth of several tens microns from the treated surface, and then there is an gradual decrease of compressive residual stress to zero [28–30]. This phenomenon may be explained by the fact that there is an absorbing layer during LSP which avoids the thermal effect from heating of the surface by the laser beam, whereas small beads with high velocity hit the metal surface resulting to the thermal effect at the top surface during shot peening.

### *6.3.2 Results of XRD Analysis*

Figure 6.6 shows X-ray diffraction (XRD) profiles of the samples with (black) and without (red) a single LSP impact. Obviously, a martensite transformation does not take place in the surface layer after LSP. In the previous works, as a consequence of near-surface plastic deformations, strain induced martensite transformation was observed after the treated AISI 304 SS by traditional surface treatment techniques, such as deep rolled, shot peening [28], and surface mechanical attrition treatment (SMAT) [31]. Accompanying plastic deformation with high strain rate, mechanical twining and shear banding become more obvious leading to occurrence of favorable conditions for formation of α-martensite at intersections of shear bands, twins or plates of ε-martensite. The amount of α-martensite in AISI 304 SS during cold rolling at room temperature was found to increase with deformation and reduction of the grain size [32].

The formation of processing-induced martensitic phases is a usual phenomenon for austenitic stainless steels, which depends on temperature, strain and strain rate [33]. It is worth noting that although the strain rate with about $10^7\ s^{-1}$ in the treated surface during LSP, the use of an absorbing layer combined with a transparent confining layer of running water avoids a thermal effect from heating

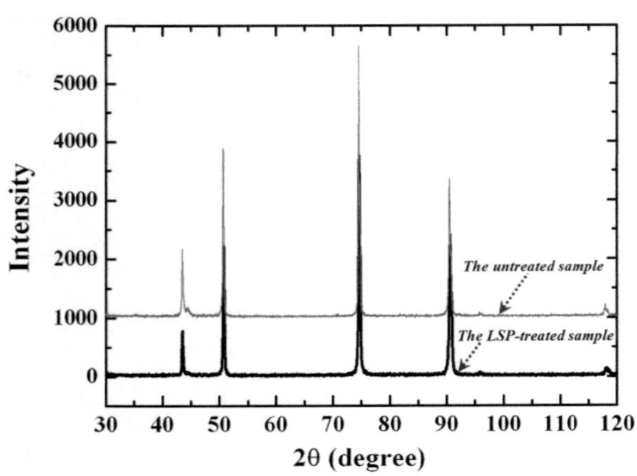


**Fig. 6.6** XRD profiles of the samples with (*black*) and without (*red*) a single LSP impact

of the surface by the laser beam, so an α-martensite transformation does not take place in the near-surface layer during LSP. In addition, Mordyuk et al. did not find the presence of deformation induced martensite in the LSPed sample without absorbing coating [34]. These phenomena need to be explained by more systematic investigations.

### 6.3.3 *Effects of LSP on Micro-Structure in the Shocked Region*

Figure 6.7 presents the OM morphologies of the cross-section of the treated sample immersed in the professional etching reagent for 30 s at room temperature. Evidence of plastic deformation is obvious in the surface layer to a depth of about 900 μm. Figure 6.7b and c are the higher magnifications of the Ellipses [A] and [B] in Fig. 6.7a, respectively. Figure 6.7b is the typical image of grains in the treated surface, while Fig. 6.7c is the typical image of grains in the substrate. It can be clearly seen from Fig. 6.7 that the average size of grains in the near-surface layer is about 2–5 μm, while the average size of the original grain in the substrate is about 7–10 μm. With the increase of the distance to the treated surface, the grain size increases, and micro-structural morphology of the plastic deformation layer subjected to a single LSP impact obviously differs from that in the substrate. As a consequence of a single LSP impact, the original grain in the near-surface layer is clearly refined. It is interesting to note that there are some grains are subdivided by cell structures which are shown as Circle C, Quadrangles D and E in Fig. 6.7b. These cell structures are the refined grains by dislocation multiplication induced by LSP, which is, in general, the reason why nano-hardness of AISI 304 SS increases.

Figure 6.8 shows the typical TEM images of different layers subjected to a single LSP impact. Figure 6.8a is the typical TEM image in the substrate, while Fig. 6.8b–d are typical TEM images in the near-surface layer after a single LSP impact. From Fig. 6.8a, it can be clearly seen that the average size of the original grain in the substrate is about 10 μm, which agrees well with the experimental results obtained in the OM observations above. From Fig. 6.8b–d, it can be seen that the original grains are subdivided by mechanical twins (MTs) into thin twin-matrix (T-M) lamellae whose width ranges from 40 to 400 nm (as seen in the Ellipses [A]–[F] in Fig. 6.8), and the MT width is about 50 nm. Dislocation lines are also observed inside twins and matrixs, but there isn't dislocation cell in these regions. Hence, deformation MT plays an important role in the refinement mechanism of LSP on AISI 304 SS.

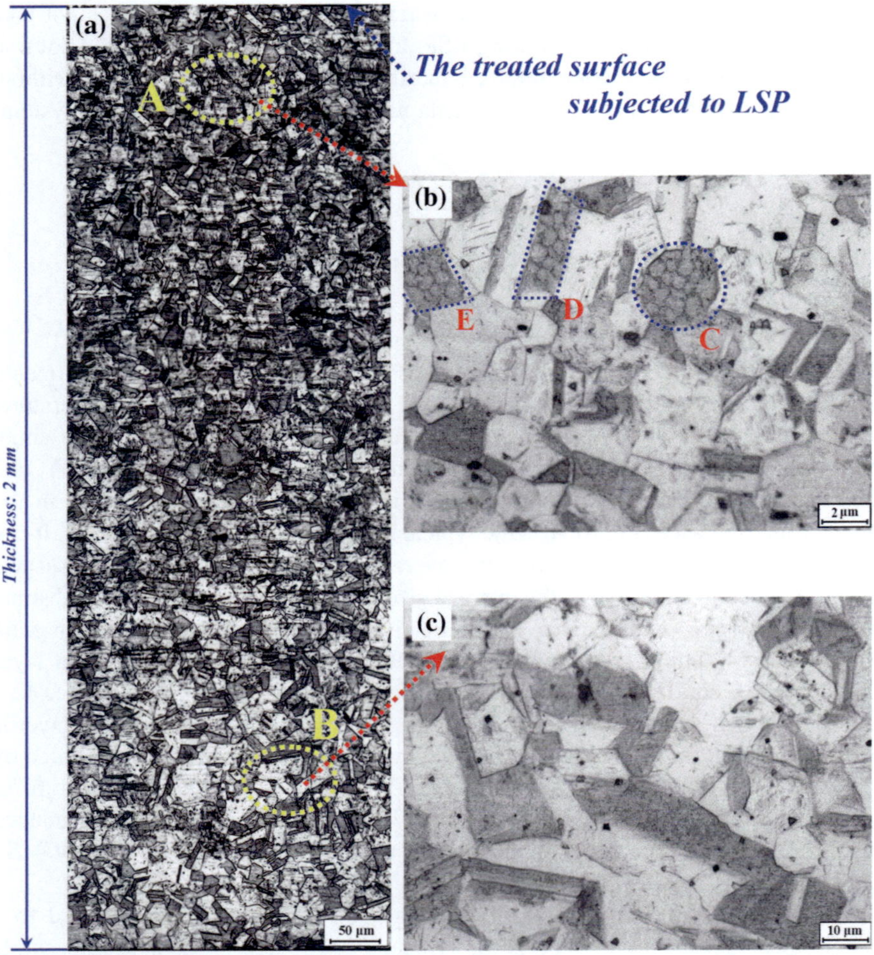


**Fig. 6.7** **a** OM morphologies of the cross-section of the treated sample immersed in the professional etching reagent for 30 s at room temperature, and **b** and **c** the higher magnifications of the *Ellipses A* and *B* in **a**, respectively

## 6.3.4 Enhancement Mechanism of LSP on Mechanical Properties of AISI 304 SS

From the above-mentioned experimental results, it can be concluded that a single LSP impact can clearly refine the original grains in the near-surface layer of AISI 304 SS mainly by MTs in a single direction. It is well known that the transformation in microstructure is accompanied by a substantial increase in mechanical properties [35]. In fact, the presence of compressive residual stresses is often the result of micro-plastic deformation accompanying the micro-structure changes [34]. So the compressive residual stress is generated in the plastic deformation

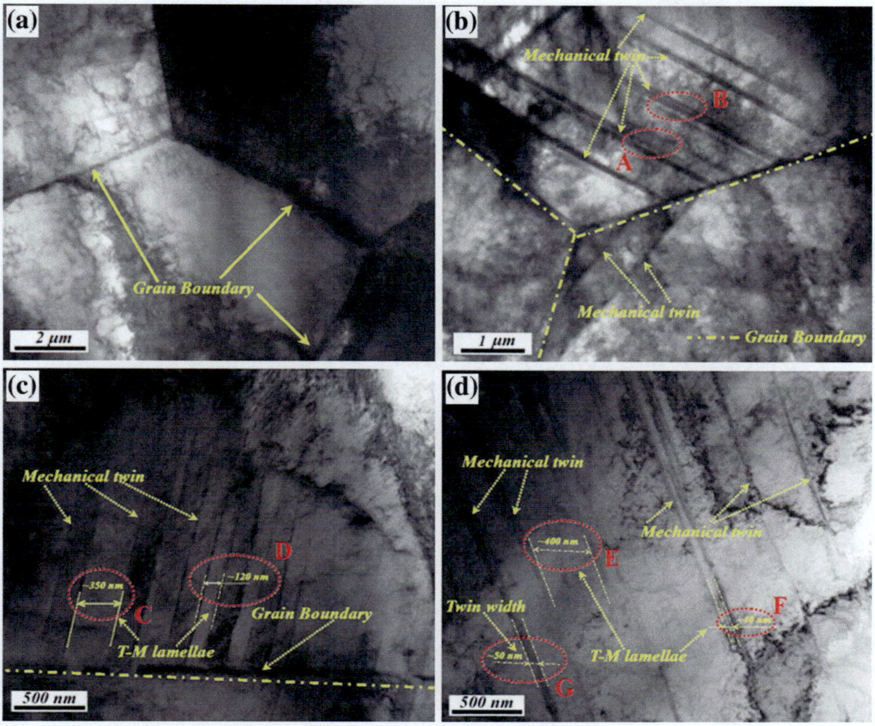


**Fig. 6.8** Typical TEM images of different layers subjected to a single LSP impact, **a** in the substrate, and **b**, **c**, and **d** in the near-surface layer

layer, which can be explained by the fact that the reaction between the laser shock wave and AISI 304 SS results in the generation of MTs and the micro-structural plastic deformation near the surface of the treated sample.

It is well known that there is a direct relation between hardness and twin spacing. Shock loading of stainless steels [36] has produced twin structures with spacing in the range of few tens to hundreds of nanometers. Hall–Petch behavior, where twin spacing leading to hardness proportional to $L^{-1/2}$, ($L$ is the twin spacing), is observed in stainless steels. Twinning-induced strengthening in austenitic stainless steel was studied by Zhang et al., and their results provided a new perspective to producing ultrahigh strength monolithic metals by utilizing growth twins with nanometer-scale spacing [37]. In addition, twin boundaries are equivalent to grain boundaries in providing Hall–Petch type strengthening where grain (or twin) boundary spacing is in the few to a few tens of micrometers range. LSP can refine the original grains in the near-surface layer by generating the MTs of AISI 304 SS, and surface nano-hardness increases with the decrease of the twin spacing.

Elastic modulus has a direct relation to fraction porosity in ultra-fine metal materials. Sanders et al. investigated the relation between the elastic modulus and porosity for Pd and Cu, and the experimental results showed that the elastic modulus decreases while the fraction porosity increases [38]. From the above-mentioned experimental results, LSP can obviously refine the original coarse grains in the LSPed region, and the local plastic deformation generated by LSP results in compact-grain structure. It can be concluded that LSP can decrease the fraction porosity of AISI 304 SS, and LSP consequently increase the elastic modulus of AISI 304 SS.

## 6.4 Mechanical Properties and Fracture Morphology of Laser Welded Joint

### *6.4.1 Mechanical Properties of Laser Welded Joint*

Figure 6.9a and b show the time-loading and tensile displacement-loading curves of the laser welded samples subjected to single-sided LSP impacts and two-sided LSP impacts, respectively. It can be found that the values of the yield strength and

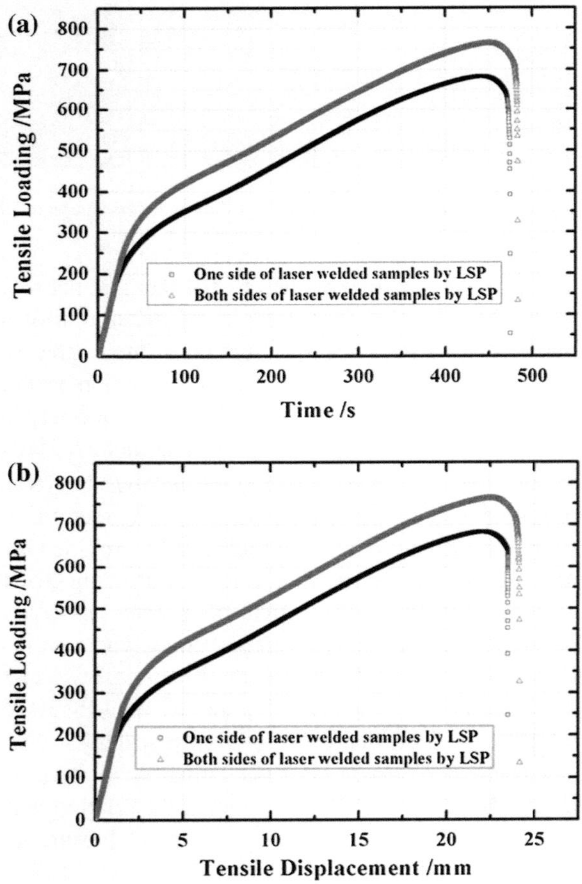


**Fig. 6.9** The *curves* of the welded samples subjected to single-sided LSP impacts and two-sided LSP impacts during the tensile test. **a** the action time-loading curves, and **b** the tensile displacement-loading curves

the tensile strength of the welded samples subjected to two-sided LSP impacts are higher than those of the welded samples subjected to single-sided LSP impacts universally.

At the same time, the mechanical properties of the welded samples subjected to different LSP shocked paths are listed in Table 6.4 in detail. The average yield strength of the welded samples subjected to single-sided LSP impacts is 240.74 MPa at 35.803 s, and the tensile strength is 683.10 MPa at 442.430 s. While the average yield strength of the welded samples subjected to two-sided LSP impacts is 318.18 MPa at 43.604 s, and the tensile strength is 763.10 MPa at 451.031 s. Compared with the welded samples treated by single-sided LSP impacts, the time when the yield strength and tensile strength appear is slightly later during two-sided LSP impacts, and the values of the yield strength and tensile strength of the welded samples treated by two-sided LSP impacts are higher. Because the laser welded AISI 304 SS with the thickness of 5 mm belonging to thick plate, and the surface crack initiation on both sides of the welded samples can be inhibited to a certain extent owing to the increasing of the compressive residual stress during the two-sided LSP impacts. Then two shock waves from reserve direction induced by two-sided LSP impacts will counteract the tensile residual stress at the mid-thickness of the samples and reduce the peak value of the tensile residual stress at the stress concentration part such as sharp-angled place [39]. Hence, the potential cracks in the inner materials and in the sharp-angled place are restrained, and the mechanical properties of the welded samples may increase at a certain degree during two-sided LSP impacts.

Figure 6.10 shows the engineering stress-strain curve of the laser welded samples during single-sided LSP impacts and two-sided LSP impacts. The whole trend of the curve is similar as that in Fig. 6.9. There is no obvious physical yield (plateau) phenomenon, and this is beneficial to the material forming. Based on Hooke Law, the elastic modulus can be calculated. During the elastic stage in Fig. 6.10, the relationship between the stress and strain of the welded samples treated by different shocked paths is linear. According to the theory, the elastic modulus of the welded samples subjected to single-sided LSP impacts is 10,829.51 MPa, and that subjected to two-sided LSP impacts is 11,689.20 MPa. In conclusion, the elastic modulus of the welded samples during two-sided LSP impacts is higher than that during single-sided LSP impacts. The material ductility can be described by the area under the stress-strain curve from the beginning of the plastic deformation to the ending of the fracture, and this can also be called static ductility. The fracture strain of two-sided LSP impacts is larger. So the ductility during two-sided LSP impacts is better than that during single-sided LSP impacts.

**Table 6.4** Tensile properties of the welded samples during LSP under different shocked paths

| Group | Yield strength /MPa | Tensile strength /MPa | yield-to-tensile strength ratio / % |
|---|---|---|---|
| Single-sided LSP impacts | 240.74 | 683.10 | 0.35 |
| Two-sided LSP impacts | 318.18 | 763.10 | 0.42 |

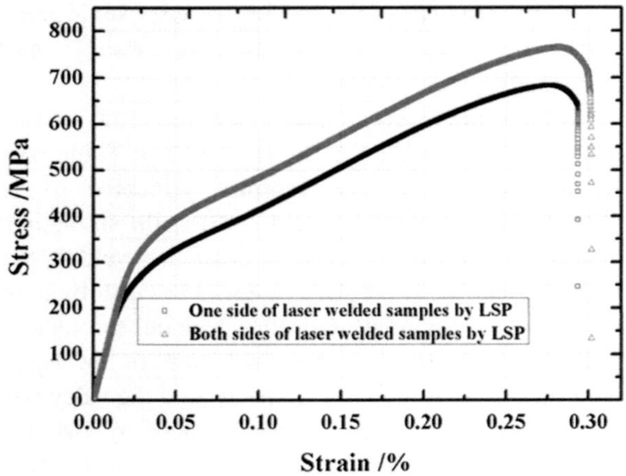


**Fig. 6.10** Engineering *stress–strain curve* of the welded samples subjected to different shocked paths

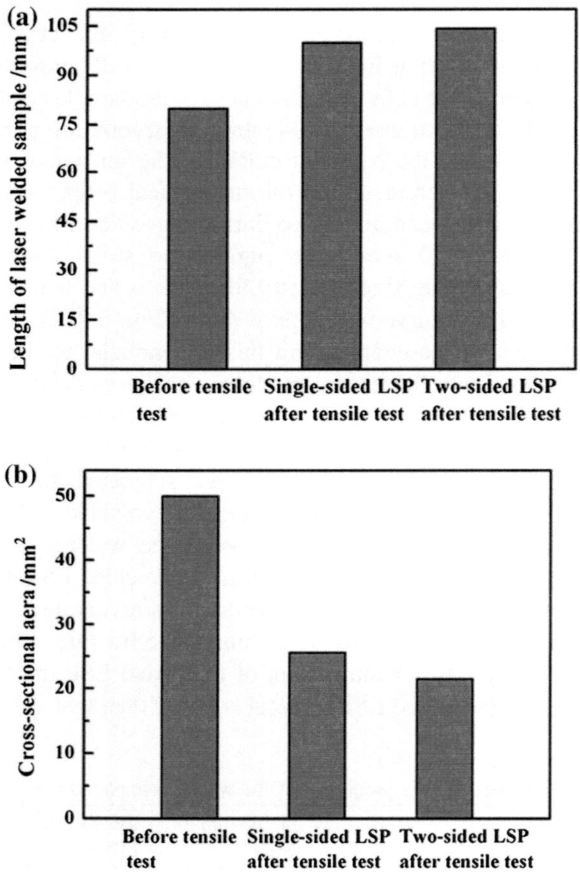


**Fig. 6.11** Comparison of the length and cross-sectional area of the welded samples subjected to different shocked paths during tensile test. **a** comparison of length, and **b** comparison of cross-sectional area

During the whole process of the plastic deformation and fracture, the capacity of the welded samples treated by two-sided LSP impacts is obviously improved to absorb the energy.

Figure 6.11 shows the comparison of the length and the cross-sectional area of the laser welded samples subjected to different shocked paths. After the tensile test, the length of the welded samples treated by single-sided LSP impacts is 100.01 mm and the cross-sectional area is 25.63 $mm^2$. The length of the welded samples treated by two-sided LSP is 104.15 mm and the cross-sectional area is 21.52 $mm^2$. According to these data, it can be figured out that the elongation and the area reduction of the welded samples treated by single-sided LSP impacts are 25.01 and 48.74 %, respectively; whereas, those treated by two-sided LSP impacts are 30.19 and 56.96 %, respectively.

### 6.4.2 Fracture Morphology of Laser Welded Joint

Figure 6.12 shows the macroscopic fracture morphology of the laser welded samples treated by different shocked paths and without LSP impacts. On the whole, the fracture morphology of the laser welded samples subjected to two-sided LSP impacts in Fig. 6.12b is the smoother than that subjected to one-sided LSP impacts in Fig. 6.12a and that without LSP impacts in Fig. 6.12c. In Fig. 6.12c, the fracture is irregular with obvious delamination splitting and some cracks. In Fig. 6.12a, there is the phenomenon of delamination splitting in the sharp corner with a few cracks around during one-sided LSP impacts. In Fig. 6.12b, the fracture subjected to two-sided LSP impacts is smooth, almost without delamination splitting. In contrast to the crack origin in the sharp-angled place by single-sided LSP impacts, the crack origin does not occur during two-sided LSP impacts. It can be said that the imperfection such as delamination splitting and cracks can be restrained by LSP impacts, and the effects of two-sided LSP impacts on the tensile fracture is better than that of single-sided LSP impacts.

Figure 6.13a–c show the fractograph of the mixed area of the laser welded joint by single-sided LSP impacts, two-sided LSP impacts and without LSP impacts, respectively. In the mixed area, there are different fracture characteristics, such as the mixing of the brittle fracture, dimple or cleavage. In Fig. 6.13a about the mixed area by single-sided LSP impacts, there is some brittle fracture facet region, as seen in the square [A] and there are a few dimples mainly in the square [B]. In the mixed area without LSP impacts, there are some huge brittle fracture facet regions, such as in the square [D] in Fig. 6.13c. A few dimples only appear in the square [E]. The ductility of the welded samples by single-sided LSP impacts in Fig. 6.13a is better than that without LSP impacts in Fig. 6.13c. As compared with Fig. 6.13a and c, more dimples in the fracture treated by two-sided LSP impacts can be found in Fig. 6.13b, such as in the square [C]. These dimples distribute more densely. A region is with the characteristics of cleavage, and the dimples are the dominant fracture mode after two-sided LSP impacts. The more the dimples,

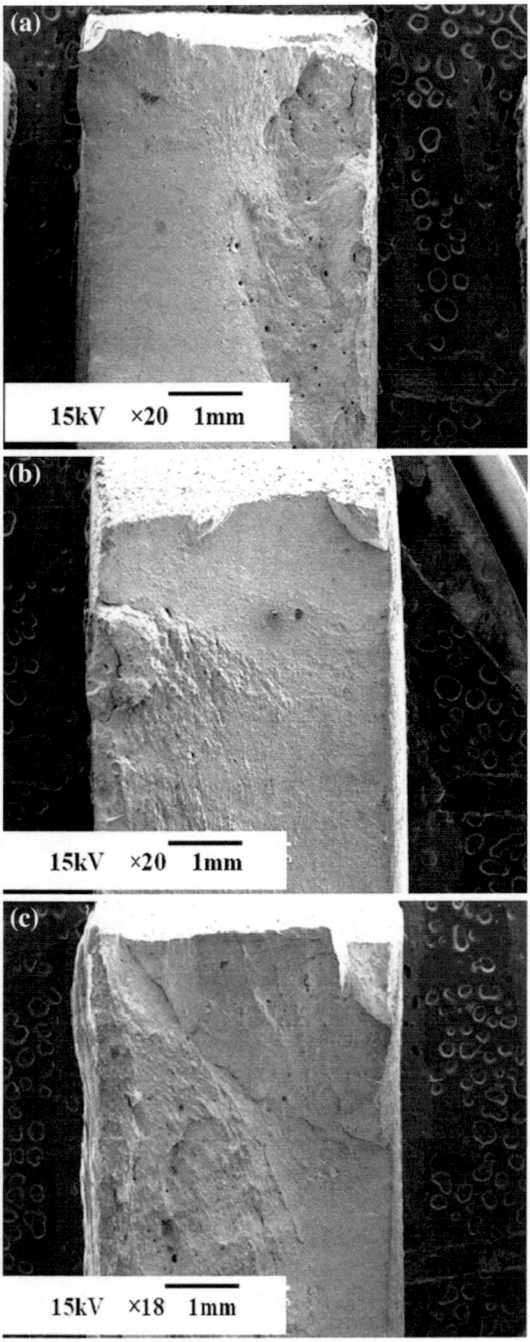


**Fig. 6.12** Typical macroscopic fracture morphology of the laser welded samples subjected to different shocked paths. **a** single-sided LSP impacts, **b** two-sided LSP impacts, and **c** without LSP impacts

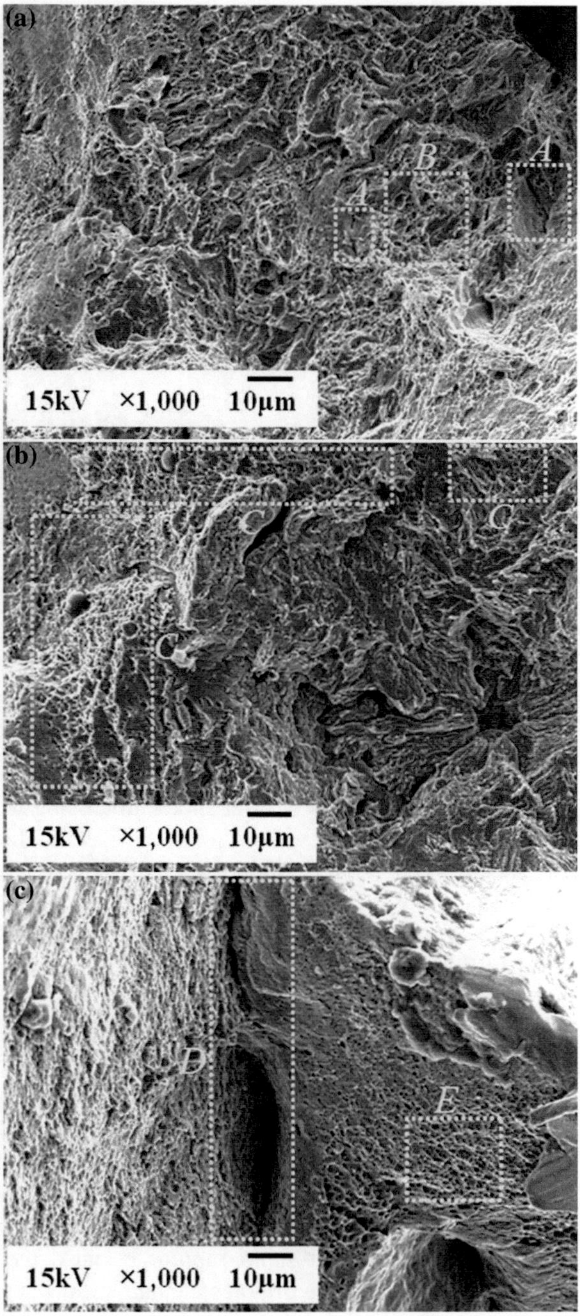


**Fig. 6.13** Typical fractograph in the mixing area of the laser welded joint processed by different shocked paths. **a** single-sided LSP impacts, **b** two-sided LSP impacts, and **c** without LSP impacts

the better the ductility is. As a whole, the ductility of the welded samples by two-sided LSP impacts in Fig. 6.13b is better than that by single-sided LSP impacts in Fig. 6.13a, even than that without LSP impacts in Fig. 6.13c. In view of it and in

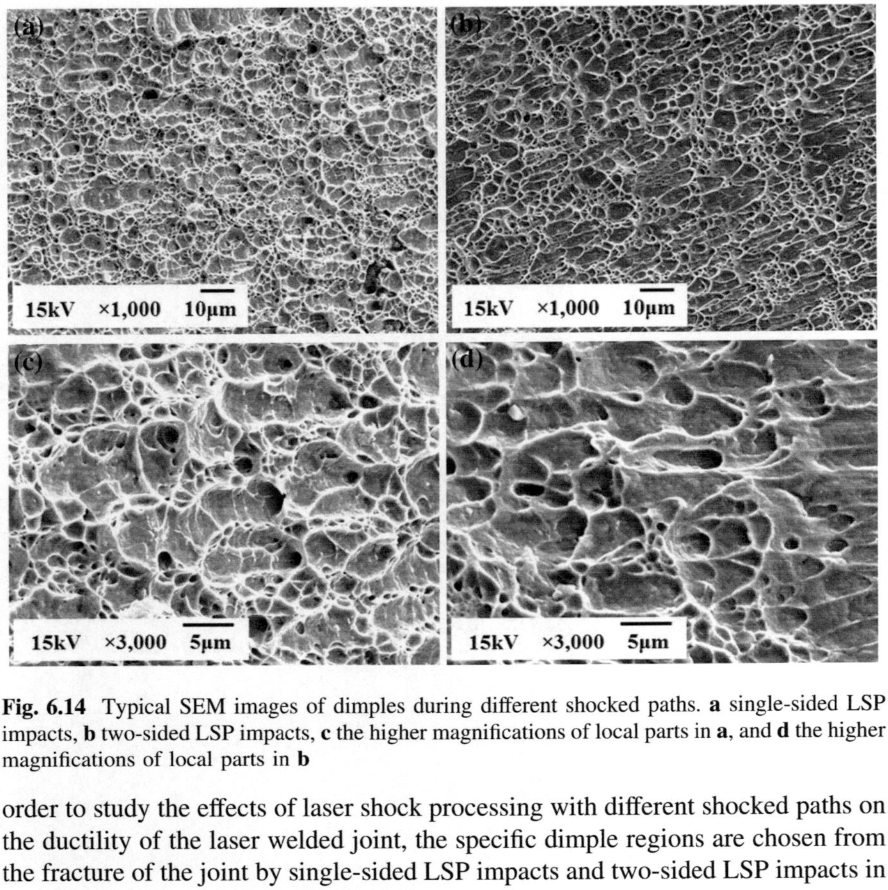


**Fig. 6.14** Typical SEM images of dimples during different shocked paths. **a** single-sided LSP impacts, **b** two-sided LSP impacts, **c** the higher magnifications of local parts in **a**, and **d** the higher magnifications of local parts in **b**

order to study the effects of laser shock processing with different shocked paths on the ductility of the laser welded joint, the specific dimple regions are chosen from the fracture of the joint by single-sided LSP impacts and two-sided LSP impacts in Fig. 6.13a and b.

Figure 6.14a and b show the dimples by single-sided LSP impacts and two-sided LSP impacts, respectively. The dimple rupture is the main ductility fracture characteristic of the welded samples after the tensile test. Compared with the equiaxed dimples during single-sided LSP impacts in Fig. 6.14a, the elongated dimples during two-sided LSP impacts in Fig. 6.14b are deeper and distributing uniformly. It can be said that the applied loadings on the tensile samples treated by two-sided LSP impacts are larger than that by single-sided LSP impacts, which changes the dimples from equiaxed shape to elongated one. Figure 6.14c and d are the higher magnifications of local parts in Fig. 6.14a and b, respectively. New small dimples occur in the big comparatively dimples in Fig. 6.14c. The fracture mode is transgranular dimple rupture because the microvoid nucleates in the grains. In Fig. 6.14d, the dimples contain shear ones, which reveals that the stain rate is increasing and the shearing mode occurs with the tensile loading. Along the direction of the tensile, the dimples grow and they are elongated. The size of the dimples is influenced by the material hardening exponent, the capacity of the plastic deformation, the stress state, the strain rate and the temperature, which is related with the magnitude of the material plastic deformation. The bigger the size, including the average diameter and the depth, of the dimples is, the better the material plastic is [40]. Relatively speaking, the dimples of the laser welded samples treated by two-sided LSP impacts correspond to this fine characteristic. In conclusion, the plastic deformation in Fig. 6.14b is more severe, and the mechanical properties such as the ductility and plastic of the laser welded samples by two-sided LSP impacts are superior to that by single-sided LSP impacts.

## 6.5 Summary

This chapter presents the effects of a single LSP impact on the nano-hardness, elastic modulus, residual stress and phase transformation of AISI 304 SS, and exhibits the mechanical properties of laser welded 304 SS subjected to different shocked paths during LSP. The following conclusions can be obtained.

1. LSP can clearly improve the nano-hardness and elastic modulus of the LSPed region in near-surface layer, and the nano-hardness and elastic modulus gradually decrease with the increase of the distance to the treated surface. The reason may be due to the fact that a single LSP impact can refine the original grain in the near-surface layer mainly by MTs in a single direction.
2. Through the tensile test, the yield strength of the laser welded samples subjected to single-sided LSP impacts is 240.74 MPa, and the tensile strength is 683.10 MPa. While the average yield strength of the welded sample subjected to two-sided LSP is 318.18 MPa, and its tensile strength is 763.10 MPa. Mechanical properties during two-sided LSP impacts have the edge over those during single-sided LSP impacts.
3. Compared with the fracture morphology of the laser welded samples by two-sided LSP impacts, the phenomenon of delamination splitting occurs with some cracks in the sharp corner of the laser welded samples by single-sided LSP impacts. The dimples are more and deeper in the two-sided LSP impacts condition. On the whole, the ductility after two-sided LSP impacts is improved than that after single-sided LSP impacts.

## References

1. Milad, M., Zreiba, N., Elhalouani, F., & Baradai, C. (2008). The effect of cold work on structure and properties of AISI 304 stainless steel. *Journal of Materials Processing Technology, 203*(1–3), 80–85.
2. Lee, H. T., & Jeng, S. L. (2001). Characteristics of dissimilar welding of alloy 690 to 304L stainless steel. *Science and Technology of Welding and Joining, 6*(4), 225–234.

3. Jamshidi, A. H., Farzadi, A., Serajzadeh, S., & Kokabi, A. H. (2009). Theoretical and experimental study of microstructures and weld pool geometry during GTAW of 304 stainless steel. *International Journal of Advanced Manufacturing Technology, 42*(11), 1043–1051.
4. Lu, S. P., Fujii, H., Nogi, K., & Sato, T. (2007). Effect of oxygen content in He-O2 shielding gas on weld shape in ultra-deep penetration TIG. *Science and Technology of Welding and Joining, 12*(8), 689–695.
5. Lima, A. S., Nascimento, A. M., Abreu, H. F. G., & Lima-Neto, P. (2005). Sensitization evaluation of the austenitic stainless steel AISI 304L, 316L, 321 and 347. *Journal of Materials Science, 40*(1), 139–144.
6. Muthupandi, V., Bala, S. P., Seshadri, S. K., & Sundaresan, S. (2003). Effect of weld metal chemistry and heat input on the structure and properties of duplex stainless steel welds. *Materials Science and Engineering A, 358*(1–2), 9–16.
7. Chen, Y. B., Lei, Z. L., Li, L. Q., & Wu, L. (2006). Experimental study on welding characteristics of $CO_2$ laser TIG hybrid welding process. *Science and Technology of Welding and Joining, 11*(4), 403–411.
8. Yan, J., Gao, M., & Zeng, X. Y. (2010). Study on microstructure and mechanical properties of 304 stainless steel joints by TIG, laser and laser-TIG hybrid welding. *Optics and Lasers in Engineering, 48*(4), 512–517.
9. Huang, Q., Hagstroem, J., Skoog, H., & Kullberg, G. (1991). Effect of laser parameter variation on sheet metal welding. *International Journal of Joining Materials, 3*(3), 79–88.
10. Padmanaban, G., & Balasubramanian, V. (2010). Optimization of laser beam welding process parameters to attain maximum tensile strength in AZ31B magnesium alloy. *Optics and Lasers in Engineering, 42*(8), 1253–1260.
11. Qi, J. F., Niu, Z., Zhang, D. Y., & Zuo, T. C. (2008). Synthesis of SiC/Ni composite coatings by laser cladding and property of erosion resistance. *Chinese Journal of Lasers, 35*(1), 297–302. (in Chinese).
12. Danny, P., & Pal, M. (2008). Q-switch Nd: YAG laser welding of AISI 304 stainless steel foils. *Materials Science and Engineering A, 486*(1–2), 680–685.
13. Park, S. H. C., Sato, Y. S., Kokawa, H., Okamoto, K., Hirano, S., & Inagaki, M. (2005). Microstructural characterisation of stir zone containing residual ferrite in friction stir welded 304 austenitic stainless steel. *Science and Technology of Welding and Joining, 10*(5), 550–556.
14. Kawahito, Y., Mizutani, M., & Katayama, S. (2009). High quality welding of stainless steel with 10 kW high power fiber laser. *Science and Technology of Welding and Joining, 14*(4), 288–294.
15. Montross, C. S., Ye, L., Wei, T., Clark, G., & Mai, Y. W. (2002). Laser shock processing and its effects on microstructure and properties of metal alloys: A review. *International Journal of Fatigue, 24*, 1021–1036.
16. Meyers, M. A., Gregori, F., Kad, B. K., Schneider, M. S., Kalantar, D. H., Remington, B. A., et al. (2003). Laser-induced shock compression of monocrystalline copper: Characterization and analysis. *Acta Materialia, 51*(5), 1211–1228.
17. Sano, Y. J., Obata, M., Kubo, T., Mukai, N., Yoda, M., Masaki, K., et al. (2006). Retardation of crack initiation and growth in austenitic stainless steels by laser peening without protective coating. *Materials Science and Engineering A, 417*(1–2), 334–340.
18. Nikitin, I., Scholtes, B., Maier, H. J., & Altenberger, I. (2004). High temperature fatigue behavior and residual stress stability of laser-shock peened and deep rolled austenitic steel AISI 304. *Scripta materialia, 50*(10), 1345–1350.
19. Nikitin, I., & Altenberger, I. (2007). Comparison of the fatigue behavior and residual stress stability of laser-shock peened and deep rolled austenitic stainless steel AISI 304 in the temperature range 25–600 °C. *Materials Science and Engineering A, 465*(1–2), 176–182.
20. Turski, M., Clitheroe, S., Evans, A. D., Rodopoulos, C., Hughes, D. J., & Withers, P. J. (2010). Engineering the residual stress state and microstructure of stainless steel with mechanical surface treatments. *Applied Physics A, 99*(3), 549–556.

21. Lu, J. Z., Luo, K. Y., Zhang, Y. K., Sun, G. F., Gu, Y. Y., Zhou, J. Z., et al. (2010). Grain refinement mechanism of multiple laser shock processing impacts on ANSI 304 stainless steel. *Acta Materialia, 58*(16), 5354–5362.
22. Oliver, W. C., & Pharr, G. M. (1992). An improved technique for determining hardness and elastic modulus using load and displacement sensing indentation experiments. *Journal of Materials Research, 7*(6), 1564–1583.
23. Wang, X. W., Wang, J. Y., Wu, P., & Zhang, H. W. (2004). The investigation of internal friction and elastic modulus in surface nanostructured materials. *Materials Science and Engineering A, 370*(1–2), 158–162.
24. San, J. F., Wang, Z. C., Li, S. H., & Liu, J. J. (2006). Nano-hardness and wear properties of C-implanted Nylon 6. *Surface and Coatings Technology, 200*(18–19), 5245–5252.
25. Bhatt, R. T., Choi, S. R., Cosgriff, L. M., Fox, D. S., & Lee, K. N. (2008). Impact resistance of uncoated SiC/SiC composites. *Materials Science and Engineering A, 476*(1–2), 20–28.
26. Chen, X., Wang, R., Yao, N., Evans, A. G., Hutchinson, J. W., & Bruce, R. W. (2003). Foreign object damage in a thermal barrier system: Mechanisms and simulations. *Materials Science and Engineering A, 352*(1–2), 221–231.
27. Hu, Y. X., & Yao, Z. Q. (2008). Overlapping rate effect on laser shock processing of 1045 steel by small spots with Nd: YAG pulsed laser. *Surface and Coatings Technology, 202*(8), 1517–1525.
28. Altenberger, I., Scholtes, B., Martin, U., & Oettel, H. (1999). Cyclic deformation and near surface microstructures of shot peened or deep rolled austenitic stainless steel AISI 304. *Materials Science and Engineering A, 264*(1–2), 1–16.
29. Wang, S. P., Li, Y. J., Yao, M., & Wang, R. Z. (1998). Compressive residual stress introduced by shot peening. *Journal of Materials Processing Technology, 73*(1–3), 64–73.
30. Lindemann, J., Buque, C., & Appel, F. (2006). Effect of shot peening on fatigue performance of a lamellar titanium aluminide alloy. *Acta Materialia, 54*(4), 155–1164.
31. Zhang, H. W., Hei, Z. K., Liu, G., Lu, J., & Lu, K. (2003). Formation of nanostructured surface layer on AISI 304 stainless steel by means of surface mechanical attrition treatment. *Acta Materialia, 51*(7), 1871–1881.
32. De, A. K., Murdock, D. C., Mataya, M. C., Speer, J. G., & Matlock, D. K. (2004). Quantitative measurement of deformation-induced martensite in 304 stainless steel by X-ray diffraction. *Scripta Materialia, 50*(12), 1445–1449.
33. Mordyuk, B. N., Milman, Y. V., Iefimov, M. O., Prokopenko, G. I., Silberschmidt, V. V., Danylenko, M. I., et al. (2008). Characterization of ultrasonically peened and laser-shock peened surface layers of AISI 321 stainless steel. *Surface and Coatings Technology, 202*(19), 4875–4883.
34. Belyakov, A., Tsuzaki, K., Miura, H., & Sakai, T. (2003). Effect of initial microstructures on grain refinement in a stainless steel by large strain deformation. *Acta Materialia, 51*(3), 847–861.
35. Suś-Ryszkowska, M., Wejrzanowski, T., Pakiea, Z., & Kurzydłowski, K.J. (2004). Microstructure of ECAP severely deformed iron and its mechanical properties. *Materials Science and Engineering: A 369*(1–2), 151–156.
36. Murr, L. E., Moin, E., & Greulich, F. (1978). The contribution of deformation twins to yield stress: The Hall-Petch law for inter-twin spacing. *Scripta Metallurgica, 12*(11), 1031–1035.
37. Zhang, X., Misra, A., Wang, H., Nastasi, M., Embury, J. D., Mitchell, T. E., et al. (2004). Nanoscale-twinning-induced strengthening in austenitic stainless steel thin films. *Applied Physics Letters, 84*(7), 1096–1098.
38. Sanders, P. G., Eastman, J. A., & Weertman, J. R. (1997). Elastic and tensile behavior of nanocrystalline copper and palladium. *Acta Materialia, 45*(10), 4019–4025.
39. Ding, K., & Ye, L. (2003). FEM simulation of two sided laser shock peening of thin sections of Ti-6Al-4 V alloy. *Surface Engineering, 19*, 127–133.
40. Ahmed, S. M., Hokkirigawa, K., & Oba, R. (1994). Fatigue failure of SUS 304 caused by vibratory cavitation erosion. *Wear, 177*(2), 129–137.

# Chapter 7
# Stress Corrosion Cracking Resistance of AISI 304 SS Subjected to Laser Shock Processing

**Abstract** This chapter investigates the effects of massive LSP impacts on the SCC behavior of AISI 304 SS and highlights the distribution of surface residual stress, the SCC initiation and micro-structure on the top surface of three types of U-bend samples. The emphasis is placed on the mechanism of massive LSP impacts on the corrosion resistance of AISI 304 SS.

## 7.1 Introduction

The shock wave can induce relatively deep compressive residual stress in the surface layer of the metals and alloys by comparison with the conventional surface treatment techniques, such as shot peening [1–6], surface mechanical attrition treatment (SMAT) [7, 8] and equal channel angular pressing (ECAP) [9, 10]. Stress corrosion cracking (SCC) is one of the most severe maintenance problems in the power generation industry today, and the investigation on SCC in power generation industry attracts more and more attentions of researchers from the explosion of the nuclear power plant in Japan. Particularly, many crack failures occur as a result of the cyclic stress combined with corrosion [11]. Austenitic stainless steel has numerous industrial applications due to a good combination of mechanical properties and corrosion resistance. However, it is extremely susceptible to localized forms of corrosion like pitting and SCC, and in particular, it is highly vulnerable to chloride SCC [12]. SCC usually occurs when the following three factors superpose simultaneously: a susceptible material, exposure to a corrosive environment, and tensile stresses above a threshold, including residual stress [13].

There have been many reports on the effects of LSP on the corrosion resistance of metallic materials. In particular, the effect has been demonstrated through actual applications as preventive maintenance against SCC in the operating nuclear power reactors. After treated by LSP, without any other protective coating, water-immersed

Y. Zhang et al., *Laser Shock Processing of FCC Metals*,
Springer Series in Materials Science 179, DOI: 10.1007/978-3-642-35674-2_7,

AISI 304 exhibits a good capability to prohibit the SCC initiation and the propagation of small pre-cracks in an environment that is more vulnerable to SCC, due to the fact that the surface residual stress was converted from tensile residual stress to the high-level compressive residual stress [14]. The influences of LSP on the pitting corrosion behavior has been investigated and evaluated, and results showed that LSP can improve the pitting corrosion behavior of 316L steel in a NaCl 0.5 M solution [15]. The SCC behavior of 316L stainless steel subjected to LSP has been investigated, and LSP can effectively prevent the initiation of SCC cracks in the boiling magnesium chloride ($MgCl_2$) solution [16]. LSP has been found to increase the pitting potential of AA 2,050-T8 aluminum alloy [17], but has no significant effect on the solubility of hydrogen in alloy 22 [18]. The above investigations focused on either the corresponding experimental results or the improvement of residual stress induced by refined micro-structures during LSP.

In fact, LSP can generate relatively deep compressive residual stress and refine the coarse grain in the surface layer of metallic materials [1, 3]. Residual stress and micro-structure of metallic materials are two important factors to restrict the SCC initiation. Few studies investigated on the effects of both residual stress and grain refinement on the SCC resistance. In industrial applications, massive LSP impacts is an effective method to induce uniform compressive residual stress across the entire surface of the metallic component, and the overlapping rate between the adjacent round spots is usually 50 % in both transverse direction and longitudinal direction [19, 20]. Compared with conventional surface treatment techniques, the influence process and the improvement mechanism of massive LSP impacts on the corrosion resistance of metallic materials are explored to a far less degree, in particular, the effects of residual stress combined with refined micro-structure during massive LSP impacts is still not well elucidated. Hence, the corrosion behavior of metallic materials subjected to massive LSP impacts is worth to be investigated.

The purpose of this chapter is to investigate the effects of massive LSP impacts on the SCC behavior of AISI 304 SS and highlight the distribution of surface residual stress, the SCC initiation and micro-structure on the top surface of three types of U-bend samples. In addition, the mechanism of massive LSP impacts on the corrosion resistance of AISI 304 SS is also revealed.

## 7.2 Experimental Procedures

### *7.2.1 Sample Preparation*

The material subjected to LSP was commercial AISI 304 SS, and its chemical composition was shown in Table 6.1. The yield strength of AISI 304 SS was 205 MPa, and its Vickers-hardness was 200 HV. All original steel plates of $15 \times 75 \times 3$ mm$^3$ were cut from the same plate, and the dimensions of the steel

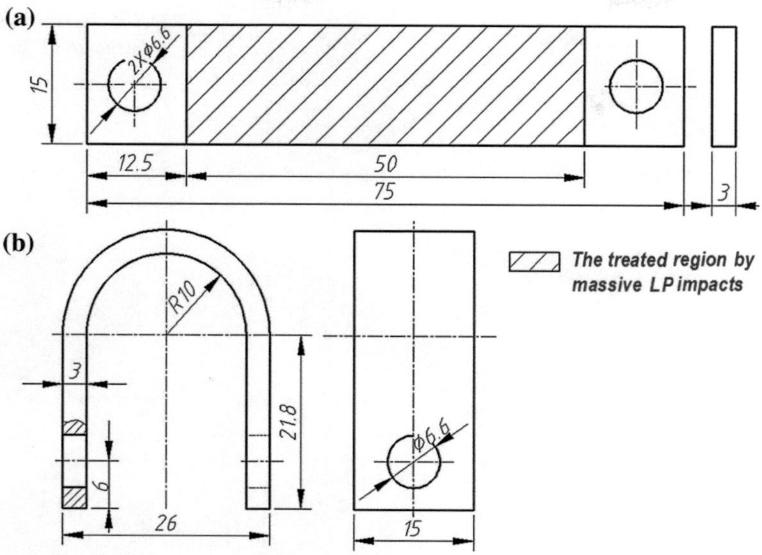


**Fig. 7.1** Dimensions of (**a**) the steel plate and (**b**) the U-bend sample (unit: mm)

plate and the U-bend sample were shown in Fig. 7.1. These steel plates were ground with different grades of SiC paper (from 500 to 1,600), and then were followed by cleaning in deionized water. Ultrasonic cleaning was used to degrease the sample surface in ethanol. Subsequently, LSP experiments were performed shortly after preparation. The as-processed steel plates were bended to the U-bend samples in accordance with ASTM G36-1994 [21]. In the present study, there are three types of U-bend samples, and each type has seven samples. Figure 7.2 shows the schematic diagrams for the three types of the U-bend samples. The first type of sample (the U-bend sample) is bended from the original steel plate, as shown in Fig. 7.2a. The second type of sample, the U-bend LSPed sample, is bended from plate after the LSP treatment, in which the LSP impacts treated surface is located in the middle region of the upper surface of the sample (Fig. 7.2b). The third one is the LSPed U-bend sample, which is made by first bending the original steel plate into U shape, and followed by LSP treatment in the middle region in the upper surface of the U-bend sample (as shown in Fig. 7.2c).

### *7.2.2 Experimental Parameters*

The massive LSP impacts were carried out using a Q-switched Nd: YAG (Neodymium doped Yttrium Aluminum Garnet) laser system operating at 1,064 nm wavelength and delivering 3 J pulse energy in 10 ns top-hat pulse, with 5 Hz repetition-rate. The laser beam was focused on the sample surface to be treated

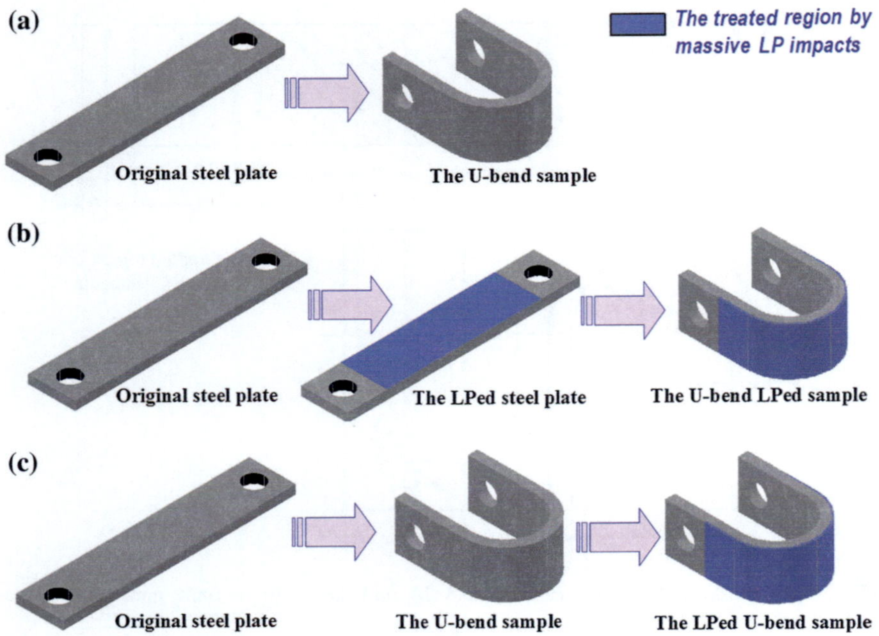


**Fig. 7.2** Schematic diagrams of three types of the U-bend samples. **a** The first type of sample (the U-bend sample): the original steel plate is bended to the U-bend sample; **b** the second type of sample (the U-bend LSPed sample): the middle region in the *upper surface* of the original steel plate is treated by massive LSP impacts and then bended to the U-bend sample; and **c** the third type of sample (the LSPed U-bend sample): the original steel plate is firstly bended to the U-bend sample, and followed by LSP treatment in the middle region in the *upper surface* of the U-bend sample

with a spot diameter of 3 mm. During LSP, all samples to be treated were submerged into a water bath, and a uniform water layer with a thickness of 1 mm was used as the transparent confining layer. The 3 M professional aluminum tape with a thickness of 100 μm (Made in USA) was used as an ablation medium for plasma initiation to protect the sample surface from the thermal damage of high-temperature plasma. During the LSP experiment, the laser beam was perpendicular to the sample surface all the time, and the overlapping rate between two adjacent spots was 50 % in order to ensure no blind area at the shocked region (as shown in Figs. 7.1 and 7.2).

The SCC tests were performed in boiling 42 % $MgCl_2$ solution, and the test solution was held at a constant boiling temperature of 143 ± 1 °C. The solution was prepared by adding a predetermined quantity of reagent grade $MgCl_2$ to distilled water into the flask. When the $MgCl_2$ solution started boiling, it was adjusted to keep the boiling temperature at 143 ± 1 °C through the addition of distilled water. All the U-bend samples were immersed into the boiling solution in the flask at the desired boiling temperature. The testing solution was changed weekly in order to maintain the same concentration during the SCC test. The failure time was recorded by considering the first time of crack initiation, along with the type of sample and the appearance of each crack. After testing, the length and feature of crack were determined on the basis of the seven measured data in the same condition.

### 7.2.3 *Measurement of Residual Stress and Micro-Structural Observations*

The XRD tests had been performed using $\sin^2\psi$ method to characterize the surface residual stresses. The X-ray beam diameter was about 2 mm, and the X-ray source was Cr-K$\alpha$ ray. The scanning starting angle and terminating angle were 145 and 153°, respectively. The diffraction plane was $\beta$ phase (311) plane. The residual stress determined using x-ray diffraction is the arithmetic average stress in a volume of material defined by the irradiated area (the spot with a diameter of 2 mm) and the depth of penetration of the x-ray beam. The measurements were repeated four times for each condition, and an average value was obtained.

After massive LSP impacts, the top surfaces of the U-bend samples used for metallographic observation were subjected to several successive steps of grinding and polishing, and then these top surfaces were etched using a professional reagent that consists of 15 cc of HCl, 10 cc of $HNO_3$, 10 cc of acetic acid, and 2/3 drops glycerine. Finally, these top surfaces were characterized by using optical microscopy (OM). At the same time, the cross-section of the samples subjected to massive LSP impacts was observed by the above-mentioned method. The magnifications of 100 and 1,000 were selected in the present work.

The micro-structures in the surface layer of AISI 304 SS with and without massive LSP impacts were characterized by using a JEM-2,100 transmission electron microscope (TEM) operated at a voltage of 200 kV. The plane-view thin foils of the surface layer were obtained first by polishing the surface layer from the surface underneath until the sample reaches about 30 μm thickness, and then perforated by electrochemical polishing (in a solution of 10 % perchloric acid and alcohol at room temperature) to make it suitable for TEM observations.

## 7.3 Results and Discussion

### 7.3.1 *High-Temperature SCC Resistance*

The SCC tests of all U-bend samples have been performed in the boiling 42 % $MgCl_2$ solution at 143 ± 1 °C. The failure times were defined by the first observation of the crack initiation. The average length of the typical crack was recorded as an important factor of SCC test.

**Table 7.1** The crack initiation time of all samples (Unit: h)

| Sample no. | 1 | 2 | 3 | 4 | 5 | 6 | 7 | Average value |
|---|---|---|---|---|---|---|---|---|
| The 1st type of sample | 17.3 | 15.7 | 16.9 | 16.1 | 15.8 | 15.4 | 15.2 | 16.06 |
| The 2nd type of sample | 113 | 119 | 102 | 108 | 115 | 107 | 109 | 110.43 |
| The 3rd type of sample | >300 | >300 | >300 | >300 | >300 | >300 | >300 | >300 |

Figure 7.3 shows the typical microscopic observations on the top surfaces of the three types of U-bend samples after SCC tests in boiling $MgCl_2$ solution. The crack initiation time of all samples is showed in detailed in Table 7.1. It can be seen from Table 7.1 that the crack nucleation can be observed on the surface of the first type of sample after immersion for an average value of 16.06 h. while the second type of sample cracks failure started after an average value of 110.43 h immersion. In contrast, there is no visible cracks can be observed on the LSPed surface for the third type of sample after immersion for a total of 300 h. A large number of cracks with a length of several hundred of micrometers are observed on the top surface of the U-bend sample (the first type of sample), as shown in Fig. 7.3a. Most of the cracks are vertical along the flanks direction, as marked by the red circles. Contrarily, the morphology on the top surface of the U-bend LSPed sample (the second type of sample) shows only one SCC crack which are branched toward the flank. More interestingly, no crack on the top surface of the third type of sample can be observed in Fig. 7.3c.

### *7.3.2 Residual Stress Distributions of Different U-Bend Samples*

The points A, B and C, as shown in Fig. 7.4, are three typical locations of the U-bend sample, which bear different residual stress. Table 7.2 shows the average values of surface residual stress at the measuring points A, B and C of all U-bend samples.

The average values of surface residual stress at the measuring points A, B and C of the first type of sample are 297, 50 and 8 MPa, while those of the second type of sample are correspondingly 130, 67 and −50 MPa, respectively. However, the corresponding points of the third type of sample are all in the state of compressive residual stress, and their average values are −364, −363 and −359 MPa, respectively. It can be seen from Table 7.2 that the surface residual stress of the original steel plates is approximately in the zero-stress state. After the original steel plate is bended to the U-bend sample, the surface residual stresses at points A and B are in a state of tensile residual stress, and their average values are 295–297 MPa. At the same time, the LSPed surface is in a state of high-level compressive residual stress, and then is converted into a state of high-level tensile

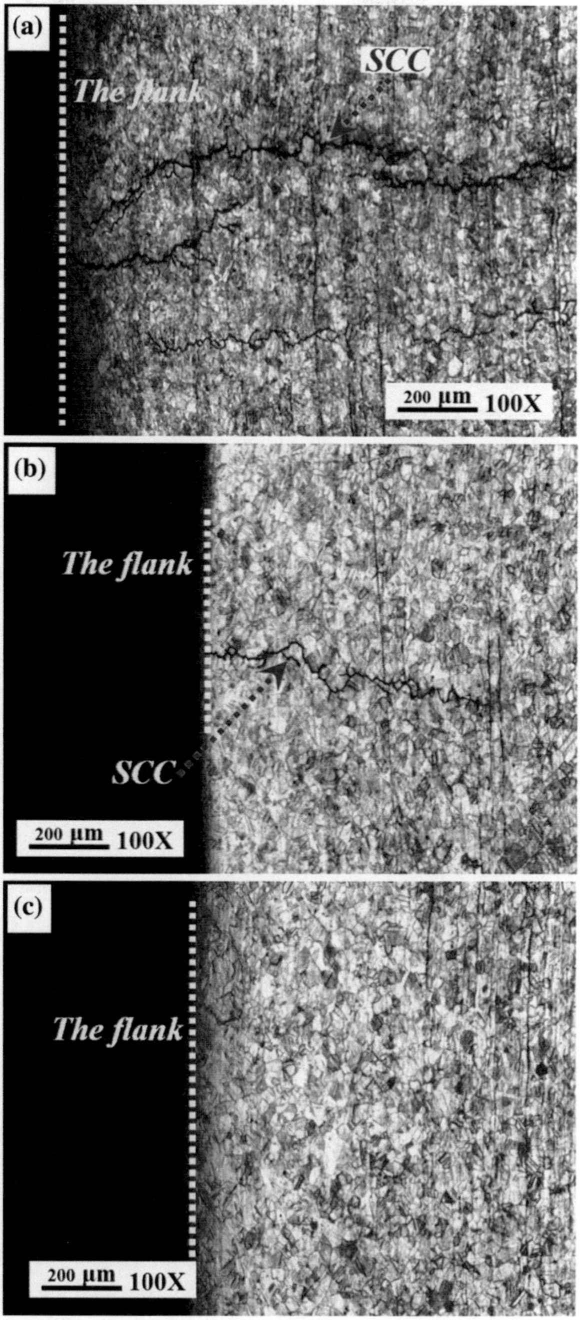


**Fig. 7.3** Typical microscopic observations on the top surfaces of three types of the U-bend samples after SCC tests in boiling $MgCl_2$ solution. **a** The first type of sample **b** the second type of sample, and **c** the third type of sample

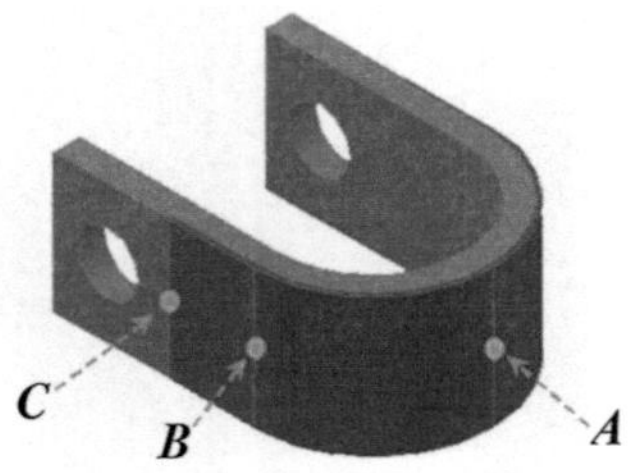


**Fig. 7.4** Schematic illustrations on the measuring points (*A*, *B* and *C*) of residual stress on the *upper surface* of U-bend sample. *Point A* the middle point of the highest ridge, *Point B* the middle point of the tangent for the plate and the upper cylindrical surface, and *Point C* the middle point of the edge line in the LSPed region

**Table 7.2** The average values of surface residual stress at different measuring points

| | State | The value of surface residual stress for measuring point (MPa) | | |
|---|---|---|---|---|
| | | A | B | C |
| The first type of sample | The original plate | 3 | 5 | 4 |
| | After the original plate is bended | 297 | 50 | 8 |
| The second type of sample | The LPed plate | −361 | −358 | −362 |
| | After the LPed plate bended | 130 | 67 | −50 |
| The third type of sample | After the original plate is bended | 295 | 53 | 7 |
| | After massive LP impacts | −364 | −363 | −359 |

residual stress when the LSPed steel plate is bended to the U-bend LSPed sample. It is interesting to find that the original steel plate is in the state of high-level tensile residual stress after bended to the U-bend sample, whereas it is in the state of high-level compressive residual stress after the middle region in the upper surface of the U-bend sample is treated by massive LSP impacts.

The above experimental results shows that high-level compressive residual stress can be induced by laser shock wave on the surface of AISI 304 SSs, which is in a good agreement with the results of our previous work [22], illustrating that that massive LSP impacts can effectively convert tensile residual stress on the surface of AISI 304 SS to relatively strong compressive residual stress. The schematic illustration of compressive residual stress induced by massive LSP impacts is shown in Fig. 7.5a. Compressive residual stress on the surface of metallic material and alloy can delay crack initiation, and compressive residual stress in the depth direction can slow the growth of micro-crack on the sample surface [14, 23]. Hence, compressive residual stress increases crack growth threshold and decreases the rate of the crack growth, consequently increasing the failure stress intensity factor during crack growth testing. Tensile residual stress is generally believed to deteriorate the SCC and fatigue failure by increasing the SCC propagation rate. Surface failure often occurs due to tensile residual stress

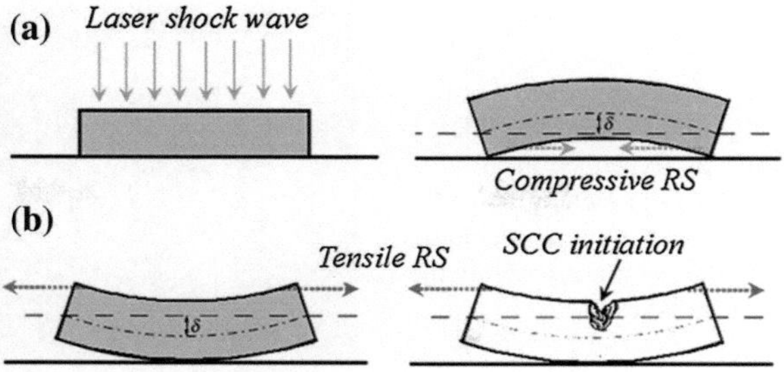


**Fig. 7.5** Schematic illustrations of **a** compressive residual stress induced massive LSP impacts, and **b** SCC initiation generated by tensile residual stress (*RS* residual stress)

developed on the outside surface of metallic material [24]. The tensile residual stress ruptures films at the crack tip and the crack grows rapidly from the bare metal exposed until the crack tip can repassivate or grow slowly to failure. The schematic illustration of SCC initiation generated by tensile residual stress is shown in Fig. 7.5b.

Massive LSP impacts can induce high-level compressive residual stress on the surface of metallic material and alloy. However, the induced high-level compressive residual stress is entirely released during the preparation process of the U-bend sample and then is converted into a state of tensile residual stress. For the second type of sample, the bending process generates tensile residual stress on the upper surface of the U-bend sample and eliminates the compressive residual stress generated by the massive LSP impacts. For the third type of sample, the middle region on the upper surface of the sample is treated by massive LSP impacts, and the LSPed region is in a state of high-level compressive residual stress. In combination with the experimental results in the Sect. 7.3.1, it can be seen that the high-level compressive residual stress on the top surface induced by massive LSP impacts can effectively retard the initiation of surface crack and thus improve the SCC resistance of AISI 304 SS.

High-level compressive residual stress induced by massive LSP impacts is one of important factors to influence the SCC resistance of AISI 304 SS. The most effective method to introduce the compressive residual stress in the U-bend sample surface is to bend the original steel plate as the U-bend sample firstly and followed by laser-peening the middle region in the upper surface of the U-bend sample.

### 7.3.3 OM Morphologies of the Cross-Section and TEM Observations of the Top Surface

Figure 7.6 shows typical OM morphologies of the cross-section of the treated sample subjected to massive LSP impacts immersed in the professional etching reagent for 30 s at room temperature. The typical image of grain in the substrate can be observed in Fig. 7.6a, while the typical image of coarse grains in the treated surface layer is shown in Fig. 7.6b. It can be clearly seen that the average size of grains in the near-surface layer is about 1–2 μm, while the average size of the

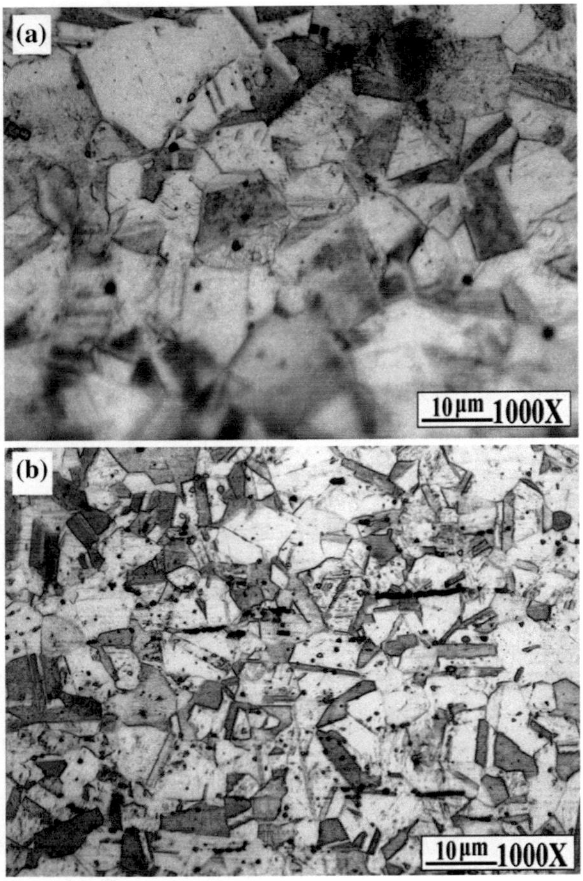


**Fig. 7.6** Typical OM morphologies of the cross-section of the treated sample subjected to massive LSP impacts immersed in the professional etching reagent for 30 s at room temperature. **a** in the substrate, and **b** in the surface layer

original grain in the substrate is about 7–15 μm. In addition, a refined grain layer with a depth of about 900 μm was observed after AISI 304 SS treated by a single LSP impact in the our previous work [22]. This has also been demonstrated by the present work that LSP can induce a deep impacted layer on the surface of AISI 304 SS (see Fig. 7.6).

Figure 7.7 shows the typical TEM micro-structures in the surface layers of two types of U-bend samples, the U-bend sample (the first type of sample) and the LSPed U-bend sample (the third type of sample). Figure 7.7a is the TEM image of a typical grain in the surface layer of the U-bend sample. It can be clearly seen that the average grain size is about 10 μm. Figures 7.7b and c are the typical TEM micro-structures on the top surface of the LSPed U-bend sample. Figure 7.7d is the SAED pattern of Circle [A] in Fig. 7.7b, and Fig. 7.7e is the high magnified image of the Circle [A] in Fig. 7.7b. Circle [A] in Fig. 7.7b and Circle [B] in Fig. 7.7c are typical mechanical twins (MTs) induced by massive LSP impacts.

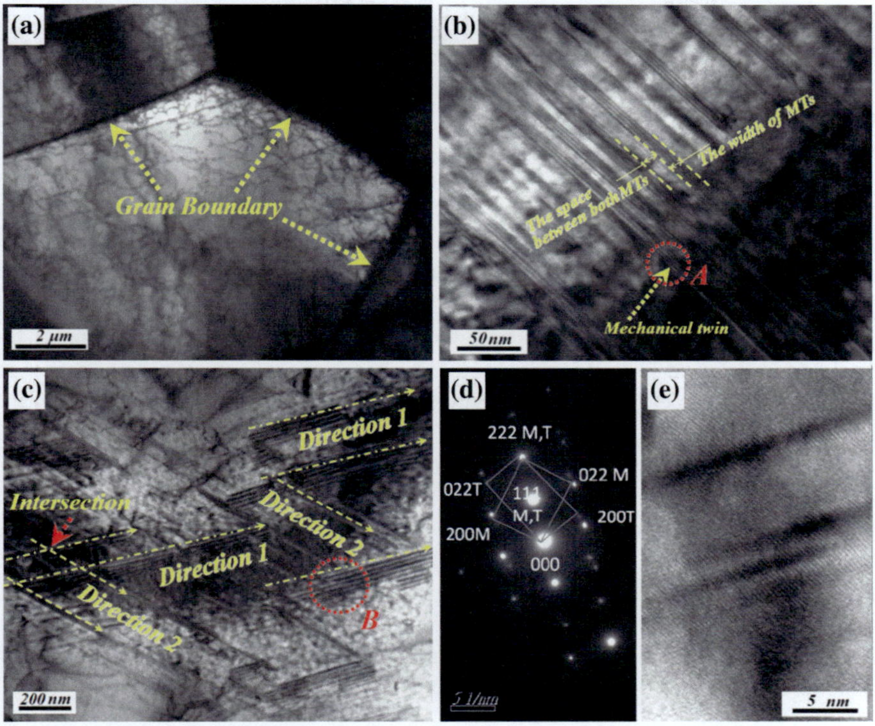


**Fig. 7.7** Typical TEM micro-structures on the *top* surfaces of different U-bend samples. **a** the first type of sample (**b**) and (**c**) the third type of sample. **d** the SAED pattern of *Circle A* in Fig. 7.7 **b**, and **e** the high magnified image of the *Circle A* in Fig. 7.7 **c**. In Figs. 7.7 **b–e**, some MTs can be clearly seen in the LSPed region. *Circle A* in Fig. 7.7 **b** and Circle *B* in Fig. 7.7 **c** are typical mechanical twins (MTs) induced by massive LSP impacts (*MT* mechanical twin)

Selected area electron diffraction (SAED) patterns taken from areas containing two adjacent lamellas indicates a diffraction patterns of two $< 011 >$ (Fig. 7.7d), and this superposition pattern are symmetrical to each other with respect to the $\{111\}$ plane, illustrating that the lamellar structure observed consists of alternate stacks of twins and matrix. Thus, we referred as this structure to twin–matrix (T-M) lamellae. From Fig. 7.7b, d and e, it can be seen that the original grains are subdivided by mechanical twins (MTs) into thin twin-matrix (T-M) lamellae whose width ranges from 10 to 60 nm, and the MT width is about 10–20 nm. Figure 7.7c shows a typical micro-structure observed at the top surface, and it can be seen that the regular intersections of MT–MT in two directions result in sub-micron rhombic blocks. Generally speaking, MTs with one direction subdivide the original coarse grains into T-M lamellae, and MTs with two directions forming MT–MT intersections divide T-M lamellae into rhombic blocks with high mis-orientations during the plastic deformation of the stainless steel. These agree well with our previous research results [22, 25]. Hence, the intersectional MTs with two directions plays an important role in the grain refinement of coarse grains in the plastic deformation layer of AISI 304 SS during massive LSP impacts. It is important to note that there are plenty of indentions at the surface of AISI 304 SS, but these indentions need to be wiped off before preparation of the TEM foils. This process may remove the refined grain layer, so the smallest grain induced by massive LSP impacts cannot be seen by using TEM.

The SCC susceptibility of metallic materials is well known to be strongly affected by their micro-structure characteristics [26]. Grain refinement can improve the mechanical properties and corrosion resistance of AISI 304 SS [27]. SCC normally initiates at the triple junction of the grain boundaries and then propagates directly along the grain boundary to another junction where the propagation is arrested. In this case, the arrested SCC has to be reinitiated for further propagation. The decrease in the overall grain size means the increase in the number of triple grain boundary junctions in the material and accordingly the probability of arresting a crack before it reaches a critical length at which failure occurs [28]. Therefore, a decrease of grain size results in an increase on the probability of crack arrest, leading to a higher resistance to crack initiation. The results provided above illustrates that massive LSP impacts can obviously refine the original grains in the surface layer of AISI 304 SS. In combination with the results in the Sect. 3.1, there are some micro-cracks on the surface of the second type of sample, and there is no crack on the surface of the third type of sample. These phenomena indicate the refined grain induced by massive LSP impacts has a beneficial effect on the SCC resistance of AISI 304 SS.

### 7.3.4 Improvement Mechanism of Massive LSP Impacts on SCC

Compared to the poor corrosion resistance performance for first type of sample and the good corrosion resistance performance for the second type of sample, all samples for the third type of sample exhibit excellent resistance to crack initiation in the boiling 42 % $MgCl_2$ solution at 143 ± 1 °C. The SCC processes of the three types of U-bend samples schematically illustrated in Fig. 7.8, and the effects of grain size and residual stress on SCC initiation are schematically illustrated in Fig. 7.9. The SCC behavior of each type sample will be discussed in terms of the experimental observations.

For the first type of sample, the original steel plate is bended to the U-bend sample, and the top surface is in a state of tensile residual stress (Step 1 in Fig. 7.8). The surface SCC initiates in a state of tensile residual stress after immersion for an average value of 16.06 h in the boiling $MgCl_2$ solution (as shown in Fig. 7.9a). For the second type of sample, the middle region of the original steel plate is treated by massive LSP impacts (Step 3 in Fig. 7.8), and then bended to the U-bend sample (Step 4 in Fig. 7.8). Although massive LSP impacts can induce compressive residual stresses on the surface of the LSPed steel plate, the bending

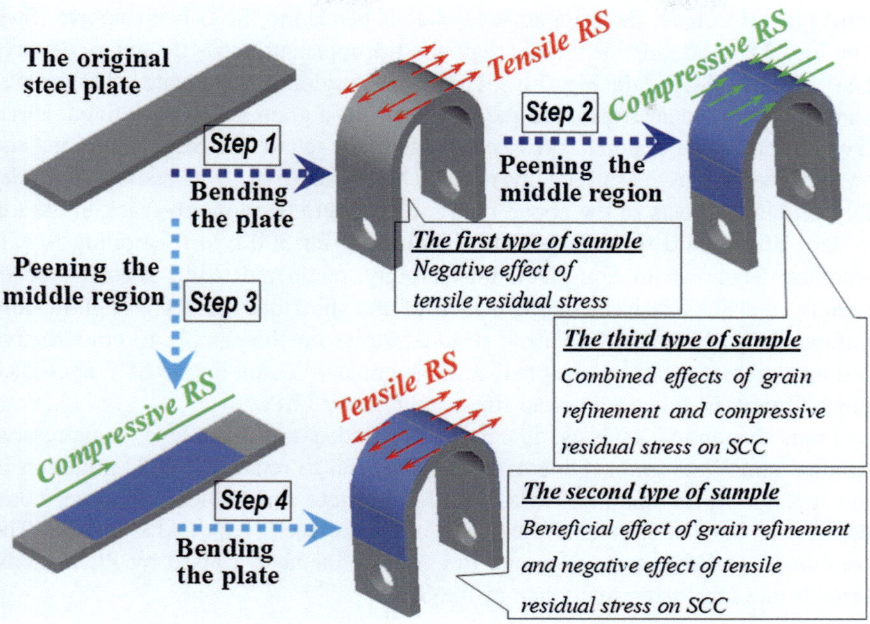


**Fig. 7.8** Schematic diagram of the SCC processes with three types (*RS* residual stress). (1) the negative effect of the first type of sample, (2) the combine effect of grain refinement and tensile residual stress of the second type of sample, and (3) the duplicate beneficial effect of grain refinement and compressive residual stress of the third type of sample

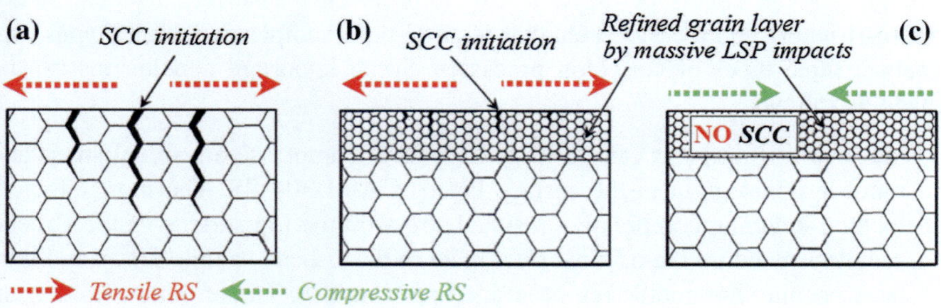


**Fig. 7.9** Schematic illustrations on combined effect of grain size and residual stress on SCC initiation of (**a**) the first type of sample (**b**) the second type of sample, and (**c**) the third type of sample (RS: residual stress)

process converts compressive residual stress into tensile residual stress on the surface of sample. During this process, massive LSP impacts refines the original coarse grains in the surface layer and retards the SCC crack initiation, but the negative effect of tensile residual stress overwhelms the beneficial effect of grain refinement during the high-temperature SCC test. Therefore, few cracks can be observed on the top surface of the second type of sample after immersion for an average of 110.43 h in the boiling $MgCl_2$ solution (as shown in Fig. 7.9b). For the third type of sample, the original steel plate is bended to the U-bend sample (Step 1 in Fig. 7.8), and then the middle region in the upper surface is treated by massive LSP impacts (Step 2 in Fig. 7.8). The LSPed region is in a state of high-level compressive residual stress, and the original coarse grain is clearly refined. High-level compressive residual stress can effectively retard the SCC initiation, and grain refinement in the surface layer is also beneficial to the SCC resistance. Under the combined effects of the above both factors, there is no micro-crack in the top surface of the third type sample after immersion for 300 h in the boiling $MgCl_2$ solution (as shown in Fig. 7.9c). Interestingly, compared to the second type of sample, the SCC resistant performance of the third one has been significantly improved by changing the tensile residual stress on the sample to compressive residual stress, thus it is plausible that the dominated reason for the SCC resistance improvement is the compressive stress induced by LSP.

From the above analysis, it can be concluded that high-level compressive residual stress and grain refinement are beneficial to retard the SCC initiation in the boiling $MgCl_2$ solution. Massive LSP impacts can induce high-level compressive residual stress and refine grains in the surface layer of AISI 304 SS. The combination of the stress state and the grain refinement, caused by PL impacts, leads to the increasing resistance of the SCC.

## 7.4 Summary

We present the effects of massive LSP impacts on the surface residual stress, micro-structure and the SCC behavior of all U-bend samples with three types, and analyze three types of corrosion processes. Some important conclusions can be made as follows:

1. Massive LSP impacts can generate high-level compressive residual stress and refine original grain in the surface layer of AISI 304 SS. The most effective method inducing compressive residual stress on the top surface of the U-bend sample is bending the original steel plate to the U-bend sample firstly and then laser peening the middle region of the upper surface. The original coarse grain on the top surface of the third type of sample is clearly refined accompanying by high-level compressive residual stresses.
2. The improvement of the SCC resistance is caused by compressive residual stress and grain refinement during LSP process. The compressive residual stress has a dominated beneficial effect on the SCC resistance, while tensile residual stresses has a negative effect on the SCC resistance. In addition, the refined grain can also effectively retard the SCC initiation.
3. After immersion in the boiling 42 % $MgCl_2$ solution at $143 \pm 1$ °C, the first type of sample cracks after an average value of 16.06 h, while the second type of sample cracks after an average value of 1,10.43 h. However, the third type of sample is tested for a total of 300 h without visible cracks in the LSPed surface,

which is attributed to the combined effects of both high-level compressive residual stress and grain refinement induced by massive LSP impacts.

## References


1. Montross, C. S., Ye, L., Wei, T., Clark, G., & Mai, Y. W. (2002). Laser shock processing and its effects on microstructure and properties of metal alloys: a review. *International Journal of Fatigue, 24*, 1021–1036.
2. Ye, C., Suslov, S., Fei, X. L., & Cheng, G. J. (2011). Bimodal nano crystallization of NiTi shape memory alloy by laser shock peening and post-deformation annealing. *Acta Materialia, 59*, 7219–7227.
3. Lu, J. Z., Luo, K. Y., Zhang, Y. K., Cui, C. Y., Sun, G. F., Zhou, J. Z., et al. (2010). Grain refinement of LY2 aluminum alloy induced by ultra-high plastic strain during multiple laser shock processing impacts. *Acta Materialia, 58*(11), 3984–3994.
4. Ye, C., Suslov, S., Kim, B. J., Stach, E. A., & Cheng, G. J. (2011). Fatigue performance improvement in AISI 4140 steel by dynamic strain aging and dynamic precipitation during warm laser shock peening. *Acta Materialia, 59*, 1014–1025.
5. Child, D. J., West, G. D., & Thomson, R. C. (2011). Assessment of surface hardening effects from shot peening on a Ni-based alloy using electron backscatter diffraction techniques. *Acta Materialia, 59*, 4825–4834.
6. Tan, L., Ren, X., Sridharan, K., & Allen, T. R. (2008). Effect of shot-peening on the oxidation of alloy 800 H exposed to supercritical water and cyclic oxidation. *Corrosion Science, 53*, 2040–2046.
7. Wen, L., Wang, Y. M., Zhou, Y., Guo, L. X., & Ouyang, J. H. (2011). Microstructure and corrosion resistance of modified 2,024 Al alloy using surface mechanical attrition treatment combined with microarc oxidation process. *Corrosion Science, 53*, 473–480.
8. Chan, H. L., Ruan, H. H., Chen, A. Y., & Lu, J. (2010). Optimization of the strain rate to achieve exceptional mechanical properties of 304 stainless steel using high speed ultrasonic surface mechanical attrition treatment. *Acta Materialia, 58*, 5086–5096.
9. Song, D., Ma, A. B., Jiang, J. H., Lin, P. H., Yang, D. H., & Fan, J. F. (2011). Corrosion behaviour of bulk ultra-fine grained AZ91D magnesium alloy fabricated by equal-channel angular pressing. *Corrosion Science, 53*, 362–373.
10. Xue, Q., Beyerlein, I. J., Alexander, D. J., & Gray, G. T. (2007). Mechanisms for initial grain refinement in OFHC copper during equal channel angular pressing. *Acta Materialia, 55*, 655–668.
11. Lu, Z. P., Shoji, T., Meng, F. J., Xue, H., Qiu, Y. B., Takeda, Y., et al. (2011). Characterization of microstructure and local deformation in 316 NG weld heat-affected zone and stress corrosion cracking in high temperature water. *Corrosion Science, 53*, 1916–1932.
12. Ghosh, S., Preet, V., Rana, S., Kain, V., Mittal, V., & Baveja, S. K. (2011). Role of residual stresses induced by industrial fabrication on stress corrosion cracking susceptibility of austenitic stainless steel. *Materials and Design, 32*, 3823–3831.
13. Mochizuki, M. (2007). Control of welding residual stress for ensuring integrity against fatigue and stress–corrosion cracking. *Nuclear Engineering and Design, 237*, 107–123.
14. Sano, Y. J., Obata, M., Kubo, T., Mukai, N., Yoda, M., Masaki, K., et al. (2006). Retardation of crack initiation and growth in austenitic stainless steels by laser peening without protective coating. *Materials Science and Engineering A, 417*(1–2), 334–340.
15. Peyre, P., Scherpereel, X., Berthe, L., Carboni, C., Fabbro, R., Béranger, G., et al. (2000). Surface modifications induced in 316L steel by laser peening and shot-peening. Influence on pitting corrosion resistance. *Materials Science and Engineering A, 280*(2), 294–302.

16. Peyre, P., Braham, C., Ledion, J., Berthe, L., & Fabbro, R. (2000). Corrosion reactivity of laser-peened steel surface. *Journal of Materials Engineering and Performance, 9*, 656–662.
17. Amar, H., Vignal, V., Krawiec, H., Josse, C., Peyre, P., Silva, S. N., et al. (2011). Influence of the microstructure and laser shock processing (LSP) on the corrosion behaviour of the AA2050-T8 aluminium alloy. *Corrosion Science, 53*, 3215–3221.
18. Marchi, C. S., Zaleski, T., Lee, S., Yang, N. Y. C., & Stuart, B. (2008). Effect of laser peening on the hydrogen compatibility of corrosion-resistant nickel alloy. *Scripta Materialia, 58*, 782–785.
19. Warren, A. W., Guo, Y. B., & Chen, S. C. (2008). Massive parallel laser shock peening: simulation, analysis, and validation. *International Journal of Fatigue, 30*, 188–197.
20. Hu, Y. X., & Yao, Z. Q. (2008). Overlapping rate effect on laser shock processing of 1045 steel by small spots with Nd:YAG pulsed laser. *Surface & Coatings Technology, 202*, 1517–1525.
21. ASTM Standard G 36-94 (2000) Standard practice for evaluating stress corrosion cracking resistance of metals and alloys in a boiling magnesium chloride solution. American Society for Testing and Materials, West Consohocken, 134–139.
22. Luo, K. Y., Lu, J. Z., Zhang, Y. K., Zhou, J. Z., Zhang, L. F., Dai, F. Z., et al. (2011). Effects of laser shock processing on mechanical properties and micro-structure of ANSI 304 stainless steel. *Materials Science and Engineering A, 528*, 4783–4788.
23. Harold, L., & Michael, R. H. (2008). The effects of laser peening on high-cycle fatigue in 7085–T7651 aluminum alloy. *Materials Science and Engineering A, 477*, 208–216.
24. Al-Obaid, Y. F. (1995). The effect of shot peening on stress corrosion cracking behaviour of 2205-duplex stainless steel. *Engineering Fracture Mechanics, 51*(1), 19–25.
25. Lu, J. Z., Luo, K. Y., Zhang, Y. K., Sun, G. F., Gu, Y. Y., Zhou, J. Z., et al. (2010). Grain refinement mechanism of multiple laser shock processing impacts on AISI 304 SS. *Acta Materialia, 58*(16), 5354–5362.
26. Kumar, S. R., Gudimetla, K., Venkatachalam, P., & Ravisankar, B. (2010). Stress corrosion cracking of Al7075 alloy processed by equal channel angular pressing. *International Journal Engineering Science and Technology, 2*, 53–61.
27. Cheung, C., & Erb, U. (1994). Application of grain boundary engineering concepts to alleviate intergranular cracking in Alloys 600 and 690. *Materials Science and Engineering A, 185*, 39–43.
28. Schino, A. D., & Kenny, J. M. (2003). Grain size dependence of the fatigue behavior of a ultrafine-grained AISI 304 stainless steel. *Materials Letters, 57*(21), 3182–3185.

# Chapter 8
# Grain Refinement of AISI 304 SS Induced by Multiple Laser Shock Processing Impacts

**Abstract** This chapter describes the micro-structural evolution and grain refinement process of AISI 304 SS subjected to multiple LSP impacts by means of cross-sectional optical microscopy and transmission electron microscopy observations, and reveals the plastic strain-induced grain refinement mechanism of FCC materials with very low stacking fault energy.

## 8.1 Introduction

Grain refinement induced by plastic strain in metals draws more and more attentions in the surface treatment field. Various techniques by plastic deformation with high strain rate, such as shot peening [1], cold rolling [2], ball milling [3], surface mechanical attrition treatment (SMAT) [4] and LSP [5], have been developed for substantially refining grains in metals and alloys in order to enhance their surface properties. Austenitic stainless steels can't be hardened by heat treatment and the greater the amount of plastic strain induced, the higher the stress is required to deform the material further. This phenomenon is known as strain hardening, which is attributed to the increasing difficulty of dislocations movement as their density increases with deformation. LSP, which is a cold machining process, is a new and promising surface treatment technique to improve the fatigue durability, corrosion, wear resistance and other mechanical properties of metals and alloys. The generated shock wave can induce sever plastic deformation (SPD) in the surface layer of the metals and alloys [6–10].

Many researchers have focused on the improvement of mechanical properties of steel subjected to LSP [11–19]. For instance, the effects of a single LSP on residual stress relaxation and hardness of the LSPed AISI 304 SS was investigated, and the results showed that LSP can improve the distributions of residual stress relaxation

Y. Zhang et al., *Laser Shock Processing of FCC Metals*,
Springer Series in Materials Science 179, DOI: 10.1007/978-3-642-35674-2_8,

and hardness [11]. The effects of LSP on the pitting corrosion resistance of 316L steel were studied, and the results showed that there were beneficial effects on the corrosion behavior after LSP [12]. The high temperature (up to 600 °C) fatigue behavior and residual stress stability of LSPed AISI 304 SS were investigated, and the results demonstrated that LSP can enhance the lifetime [13]. LSP without protective coating has been applied to water-immersed AISI 304 and AISI 316L SS, and it can effectively retard crack initiation and growth in austenitic stainless steels due to the fact that the surface residual stress of both materials was converted from tensile stress to compressive stress of several hundred MPa [14]. The distribution of micro-hardness and micro-structural morphology for ANSI 321 SS in depth direction were investigated after LSP, and the formation mechanism of the dislocation-cell structure in such stainless steels was established [15]. Besides, the relationship between the micro-hardness and the average dislocation density of the stainless steel by LSP was established [9]. The above researches have focused on the improvement of residual stress, micro-hardness, fatigue life, and corrosion resistance of austenitic stainless steel due to a single LSP impact. In fact, multiple LSP impacts may be a more effective method to significantly increase the value of compressive residual stress and fatigue life of metal components. The influences of the LSP with repeated impacts on the residual stress distribution of steel were analyzed and simulated by experiment [16] and finite element method [17], respectively. A number of researches show that the improvement of mechanical properties and fatigue life of the stainless steel is directly related to the generation of the dislocation and the micro-structural deformation near the surface during LSP process [18].

Mechanisms for the micro-structure evolution under multiple LSP impacts are of great practical importance because of the close relationship between the micro-structure and mechanical property of metal material. The grain refinement mechanism of the LY2 Al alloy with high stacking fault energy (SFE) (about 142 $mJ/m^2$) induced by ultra-high plastic strain during multiple LSP impacts was systematically investigated in our previous work [19]. The grain refinement mechanism of multiple LSP impacts on the alloy with low SFE lags well behind compared with those alloys with high SFE, and the interaction between shock waves and the resultant micro-structural changes during multiple LSP impacts is still pending. So the grain refinement mechanism of multiple LSP impacts on the alloys with low SFE is worth investigation.

The aim of this chapter is to investigate the effects of the multiple LSP impacts on the change of micro-structure in different layers of the treated sample along depth direction. The underlying grain refinement mechanism of plastic deformation after multiple LSP impacts on AISI 304 SS with a low SFE (about 17 $mJ/m^2$) is entirely revealed.

## 8.2 Experimental Procedures

### *8.2.1 Principle and Experimental Procedure of LSP*

The LSP process utilizes high energy laser pulses (several GW/cm$^2$) fired at the surface of a metal covered by two layers, namely an absorbing layer and a water confining layer. The absorbent material vaporizes and forms a plasma when a laser pulse with sufficient intensity passes through the transparent confining layer and hits the surface of the material. The plasma continues to strongly absorb the laser energy until the end of the energy deposition. The rapidly expanding plasma is trapped between the sample and the transparent confining layer, creating a high surface pressure, which propagates into the material as a shockwave [20]. When the pressure of the shockwave exceeds the dynamic yield strength of the metal, it produces plastic deformation in the near-surface of the metal [21]. The LSP principle is schematically shown in the literature [19].

### *8.2.2 Experimental Material and Processing Parameters*

The samples manufactured by AISI 304 SS were cut into a rectangle shape with dimensions of $10 \times 10 \times 2$ mm$^3$ (width × length × thickness). The chemical composition and mechanical properties of AISI 304 SS were shown in Tables 7.1 and 7.2, respectively. Prior to the LSP treatment, the sample surfaces were polished with SiC paper with different grades of roughness (from 500 to 2,400), followed by cleaning in deionized water. Ultrasound in ethanol was used to degrease the sample surface, and LSP experiments were conducted shortly after preparation.

The LSP experiments were performed using a Q-switched Nd: glass laser operating at 0.5 Hz repetition-rate with a wavelength of 1,054 nm and the FWHM of the pulses was about 20 ns. The spot diameter was 2 mm. Samples were submerged into a water bath when they were processed by LSP. A water layer with a thickness of about 1 mm was used as the transparent confining layer and the professional aluminum tape with a thickness of 100 μm was used as an absorbing layer to protect the sample surface from the thermal effect. Laser energy was 5 J. The processing parameters used in LSP were shown in Table 7.3 in detail. During multiple LSP impacts, the laser beam is perpendicular to the sample surface, and it is kept the multi-laser effect at the same location of the sample, and the aluminum tape was replaced after each impact during multiple LSP impacts.

### 8.2.3 Micro-Structural Observations

After multiple LSP impacts, the samples used for metallographic investigation were cut as the sections perpendicular to the sample surface, and then subjected to several successive steps of grinding and polishing. After that, the vertical sections of the samples were etched using a professional reagent that consists of 15 cc of HCl, 10 cc of $HNO_3$, 10 cc of acetic acid, and 2/3 drops glycerine, and then characterized by cross-sectional OM observations.

The micro-structural evolutions of the different layers in the treated samples subjected to different LSP impacts (1–3) was characterized by using a JEM-2100 TEM operated at a voltage of 200 kV. Cross-sectional thin foils for TEM observations were prepared in the following steps: (1) cutoff two pieces ($1.5 \times 1.5 \times 10$ mm$^3$ in size) of the treated sample and bond them face-to-face together; (2) put it into a 3 mm diameter copper tube and bonded them together; (3) grind it carefully to about 30 μm in thickness; (4) dimple and ion-thin it to perforation at room temperature.

For the sake of convenience, the treated samples subjected to two LSP impacts are taken as the samples subjected to multiple LSP impacts in Sect. 7.3. The cross-sections of the treated samples by LSP are observed by optical microscopy (OM) and transmission electron microscope (TEM).

## 8.3 Grain Size Variation Along Depth Direction

After multiple LSP impacts, as the strain decreases from maximum (up to $10^7$ s$^{-1}$) at the top surface to zero in the substrate, the structure evolution process may be signed by the micro-structure characteristics (with different strain rates and strains) at different depths. Detailed cross-sectional micro-structural observations are needed in order to systematically understand the in-depth micro-structures developed in different layers of the samples during multiple LSP impacts.

The cross-sectional observation of AISI 304 SS sample by two LSP impacts in Fig. 8.1, which presents the OM morphologies of the cross-sections of the samples in the SPD layer and the minor plastic deformation (MPD) layer immersed in the professional etching reagent for 30 s at room temperature. It can be clearly seen from Fig. 8.1 that the thickness of the SPD layer subjected to two LSP impacts is about 20 μm. Figure 8.2a, b show the TEM images of grains in the surface layer and the substrate, respectively. From Figs. 8.1 and 8.2, it can be clearly seen that the average grain size in the SPD layer is about 1–2 μm, while the average size of the original grain in the substrate is about 7–10 μm. With an increasing depth from the top surface, the grain size increases, and micro-structure morphology of the SPD layer subjected to two LSP impacts obviously differs from that in the substrate. As a consequence, the repeated LSP impacts at ultra-high strain rate onto the sample surface leads to SPD in the surface layer.

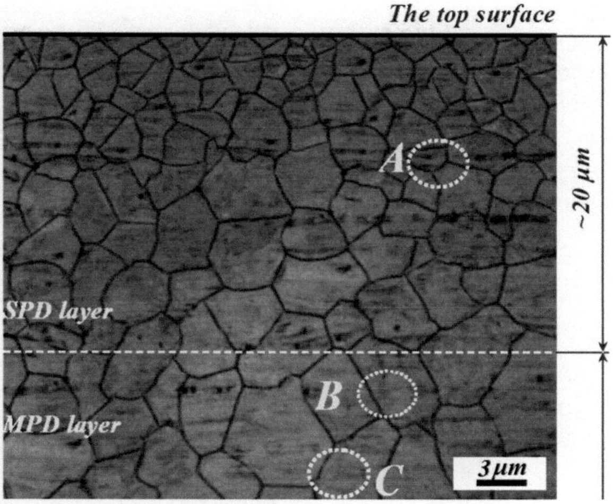


**Fig. 8.1** Typical OM morphology of the cross-section immersed in the professional etching reagent for 30 s at room temperature

## 8.4 Micro-Structural Evolvement Along Depth Direction

The TEM observations in the SPD layer, the MPD layer and the substrate subjected to two LSP impacts in Fig. 8.3, in which five typical deformation-induced micro-structure features are identified: mechanical twin (MT), planar dislocation array (PDA), stacking fault (SF), dislocation line (DL) and dislocation tangle (DT).

Figure 8.3a–c show the TEM images of the top surface subjected to two LSP impacts. Figure 8.3b is the magnified image of the *Quadrangle C* in Fig. 8.3a, and it shows a typical micro-structure observed at the top surface, and it can be seen that the intersections of MT-MT in two directions (as shown in the Ellipse [A]) result in submicron rhombic blocks. Figure 8.3c is the magnified image of the *Ellipse B* in Fig. 8.3b and it can be seen that the average dimension of the submicron rhombic block is about 60 × 120 nm (as shown in the Rhombus [A]). In Fig. 8.3c, there are also regular intersections of MTs in two directions at the top surface. Figure 8.3d shows the TEM images of the SPD layer at the depth of 10 μm from the top surface which is shown as the Ellipse [A] in Fig. 8.1, and it can be seen that there are many MTs in one direction (Direction 1) and many PDAs in the other direction (Direction 2). Intersections of MT-PDA in two directions result in micron rhombus whose width ranges from 1 to 3 μm. Figure 8.3e and f show the TEM images of the MPD layer at the depth of 25 μm from the top surface, which is shown as the Ellipse [B] in Figs. 8.1, and 8.3f is the high magnified image of the Ellipse [A] in Fig. 8.3e. In this region, micro-structure is characterized by PDAs formed on different slip planes, and it can be seen from Fig. 8.3e and f that there are many intersections of PDA-PDA in two directions. These intersections of PDA-PDA in two directions result also in submicron rhombus, and the spacing between PDAs ranges from 600 nm to 1 μm. Furthermore, there are some SFs in some grains (as shown in Fig. 8.3e and f). Figure 8.3g is the TEM image of the micro-structure in the substrate, which shows the typical

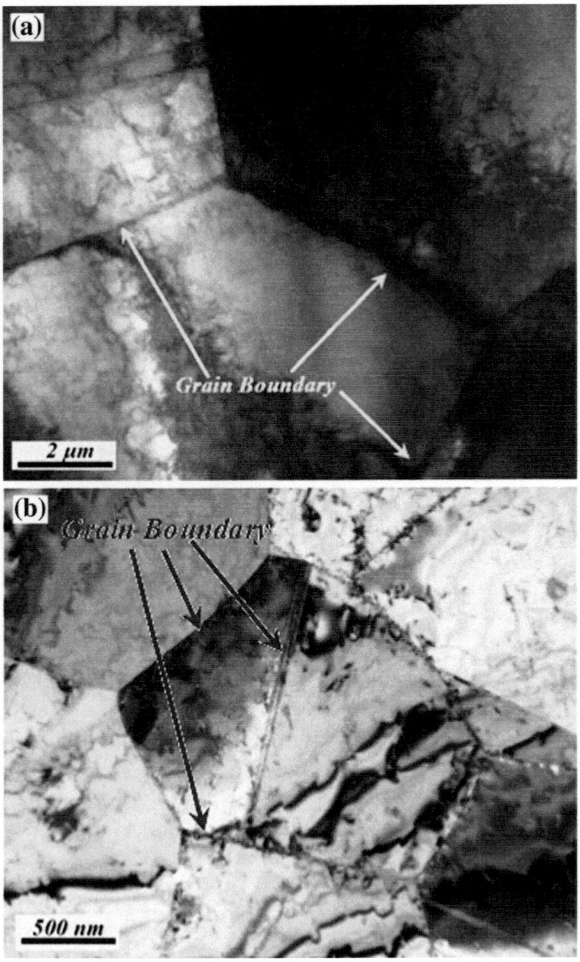


**Fig. 8.2** TEM images of grains subjected to two LSP impacts. **a** in the substrate, and **b** in the surface layer

TEM image (as shown in the Ellipse [C] in Fig. 8.1) of the MPD layer at the depth of 30 μm from the top surface. It can be seen that there are plenty of SFs in two intersectional directions and DLs in multiple directions. From Fig. 8.3a–g, it can be seen that the intersectional angle between the micro-structures (such as MT, PDA, SF) with two directions is about 70.5, and the intersectional micro-structures between two directions subdivide the coarse grain into submicron rhombuses. In addition, high densities of DTs and DLs are still visible in these micro-structure grids.

It can be concluded from the above experimental results that there are plenty of MT-MT (or MT-PDA) intersections in the SPD layer and the PDA-PDA (or PDA-SF) intersections in the MPD layer after two LSP impacts, indicating the change of micro-structures with an increase of depth from the top surface.

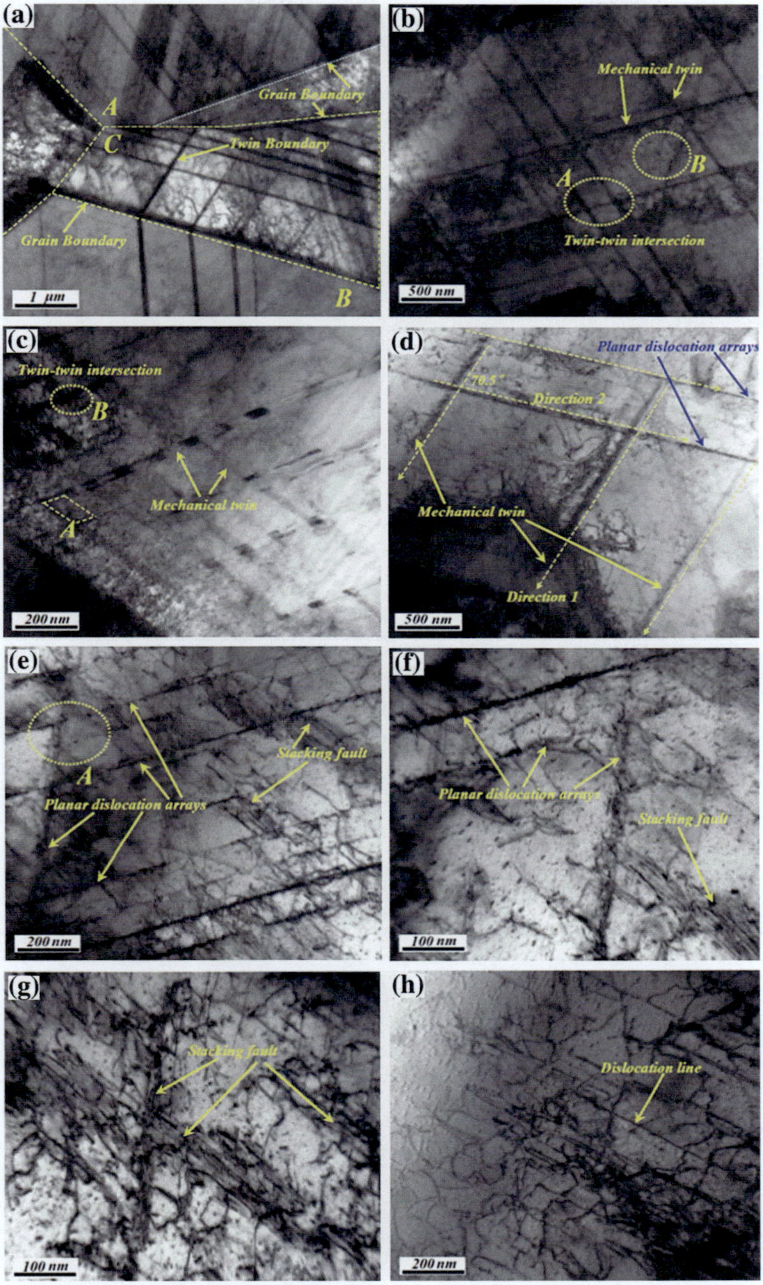


**Fig. 8.3** Typical TEM images of different layers subjected to two LSP impacts. (**a**) at the *top* surface in the SPD layer (**b**) the magnified image of the *Quadrangle C* in the (**a**, **c**) the high magnified image of the *Ellipse B* in (**b**, **d**) in the SPD layer at the depth of 10 μm from the *top* surface (**e**) in the MPD layer at the depth of 25 μm from the *top* surface (**f**) the magnified image of the *Ellipse A* in (**d**, **g**) in the MPD layer at the depth of 30 μm from the *top* surface, and (**h**) in the substrate

## 8.5 TEM Observation of the Top Surface

Figure 8.4a and b show the typical TEM images of the top surface subjected to a single LSP impact, and Fig. 8.4b is the high magnified image of the Ellipse [A] in the Fig. 8.4a. It can been seen that the original coarse grains are subdivided by MTs, whose width is about 10–30 nm, into thin twin-matrix (T-M) lamellae whose

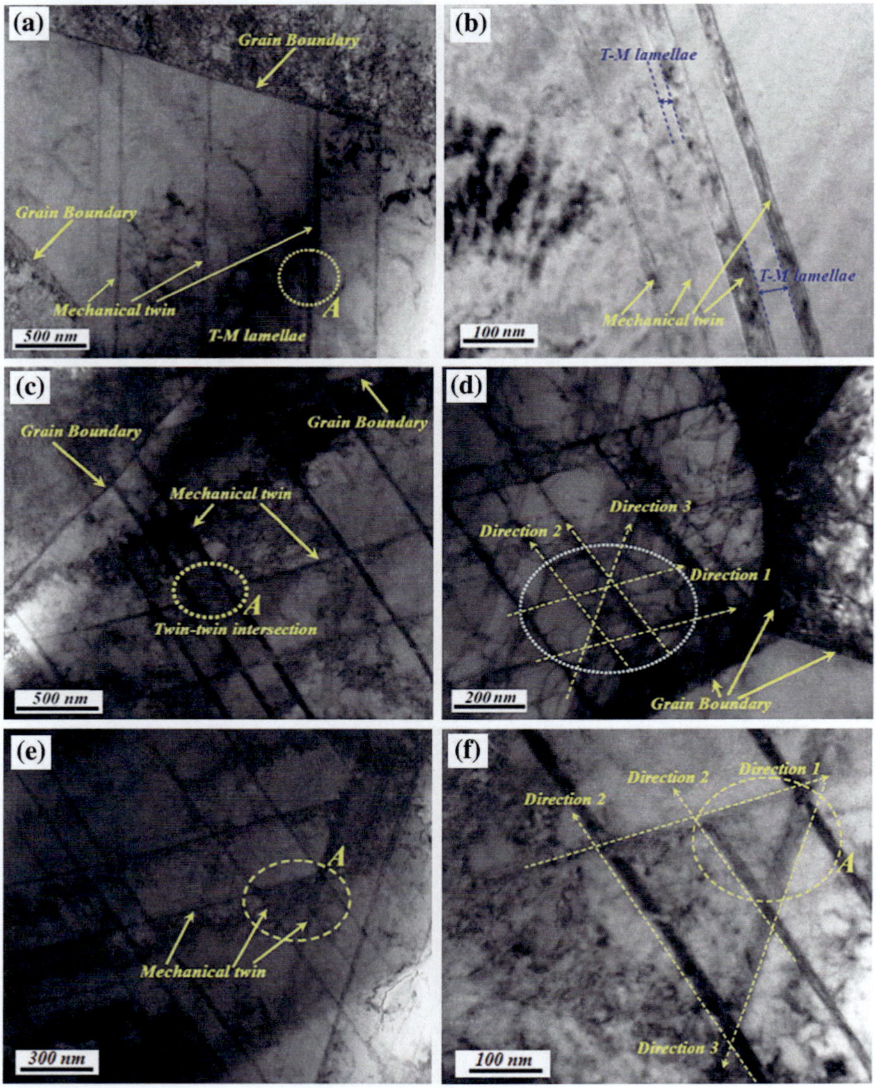


**Fig. 8.4** Typical TEM images of the top surface subjected to different LSP impact times. (**a**) a single LSP impact (**b**) the magnified image of the *Ellipse A* in the (**a**, **c**) two LSP impacts (**d**) and (**e**) three LSP impacts, and (**f**) the magnified image of the *Ellipse A* in the (**e**)

width ranges from 40 to 700 nm, as can be seen in Fig. 8.4b. DLs and DTs are also observed inside twins and matrixs, but there isn't dislocation cell in these regions. Figure 8.4c is the TEM image of the top surface subjected to two LSP impacts, shows a typical micro-structure observed at the top surface, and it can be seen that the regular intersections of MT-MT in two directions result in submicron rhombic blocks. The detailed description of the TEM image of the top surface subjected to two LSP impacts is shown in Sect. 7.3. Figure 8.4d–f show the typical TEM images of the top surface subjected to three LSP impacts, and Fig. 8.4f is the high magnified image of the Ellipse [A] in the Fig. 8.4e. It can be clearly seen from Fig. 8.4d and f that there are plenty of MT-MT intersections in three directions. During the third LSP impact, the submicron rhombic blocks are refined by the MTs in the third direction, which results in submicron triangular blocks whose length is ranges from 100 to 200 nm, as can be clearly seen in Fig. 8.4f.

## 8.6 Micro-Structural Evolution Process of the Top Surface

The strain and strain rate decrease with the increasing depth from the top surface by multiple LSP impacts, and various micro-activities are normally motivated, including MTs, PDAs, SFs, DTs and DLs in the f.c.c. materials (AISI 304 SS) with low SFE.

Unlike the grain refinement process via dislocation activities in the subsurface layer [19, 22, 23], MTs play an important role in grain refinement in the top surface at ultra-high strain rate in the SS sample by multiple LSP impacts. The grain refinement mechanism at the top surface by multiple LSP impacts can be schematically illustrated in Fig. 8.5, in which each state will be discussed in terms of the experimental observations.

During the first LSP impact, MT boundaries are parallel to each other and they subdivide the coarse grain into thin T-M lamellae at the top surface, whose width ranges from 40 to 700 nm (state (I) in Fig. 8.5), as can be clearly seen in Fig. 8.4 a and b. After the second LSP impact, the regular intersections of MT-MT in two directions result in submicron rhombic blocks with dimension of about $60 \times 120$ nm (state (II) in Fig. 8.5), as can be clearly seen in Figs. 8.3b and 8.4c. After the third LSP impacts, the submicron rhombic blocks are refined by the MTs in the third direction, which results in submicron triangular blocks whose length is ranges from 100 to 200 nm (state (III) in Fig. 8.5), as can be clearly seen in Fig. 8.4d and f. Due to the ultra-high strain rate at the top surface subjected to multiple LSP impacts, these MTs of submicron subdivided blocks turn gradually into subgrain boundaries, which eventually transform into equiaxed refined grains by development of subgrain boundaries. It is noted that the continuous dynamic recrystallization (DRX) may take place (state (IV) in Fig. 8.5) in this process, that is to say, the DRX processes result in a progressive accumulation of boundary misorientation, and finally lead to a gradual transition of boundaries character until the formation of high angle grain boundaries [24].

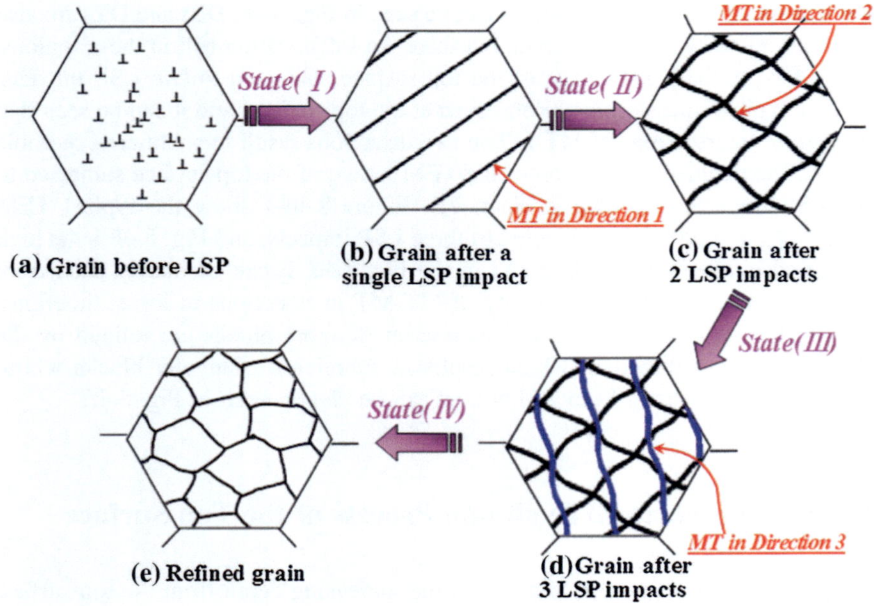


**Fig. 8.5** Schematic illustration showing micro-structural evolution process of the *top* surface induced by different LSP impact times

Generally, MTs with one direction subdivide the original 3-dimensional coarse grains into submicron 2-dimensional T-M lamellae, and MT-MT intersections divide two sets of T-M lamellae into rhombic blocks with high misorientations during the plastic deformation of the stainless steel, so the deformation twinning (instead of dislocation slip) becomes the preferred mode of plastic deformation in the f.c.c. materials [25–27]. It is anticipated that with increasing strain, multiplication of deformation twins induces more MT-MT intersections, which refine the structure into irregularly shaped grains with large misorientations [25]. In the present work, these findings provide a strong evidence to validate the above argument that the refined structures originate from subdividing the submicron rhombic blocks by MTs in the third direction into submicron triangular blocks (as shown in Figs. 8.5d and 8.4f). The previously reported approaches of grain refinement by SPD all seemed to fail to obtain refined grains with triangular blocks. The difference between LSP and the previously reported SPD processes, e.g. ball milling [3], SMAT [4, 22, 25, 27], equal channel angular pressing [28] and high pressure torsion [29], may lie in the strain rate. In LSP, the strain rate is estimated to be about $10^7\ s^{-1}$ with ultra-short laser pulses, while the strain rate is about $10^3\ s^{-1}$ in the top surface of the treated sample during ball milling and SMAT, and that is $10–10^2\ s^{-1}$ in the top surface of the treated sample during high pressure torsion and ball rolling [3, 29]. As presented in a paper about refining copper by using SMAT [30], strain rate plays a key role in refining grains. This phenomenon can be attributed to the ultra-high strain rate with ultra-short laser pulse (in nanosecond order) during multiple LSP impacts. Hence, the ultra-high strain rate with ultra-short laser pulse plays a key role in the formation of refined grains during plastic deformation of the 304 SS subjected to multiple LSP impacts.

In terms of the experimental observations and analysis, the grain refinement mechanism at the top surface of AISI 304 SS as a function of LSP impact time can be summarized as follows: (1) high-density micro-MTs divide the initial coarse grains into micrometer-thick T-M lamellae; (2) MTs with multiple directions further subdivide the T-M lamellae into equiaxed micro-sized irregularly shaped blocks; (3) the subdivided blocks evolve into randomly refined grains.

## 8.7 Micro-Structural Evolution Along Depth Direction

The intersectional micro-structure with multiple directions plays an important role in the grain refinement of coarse grains in the plastic deformation layer of the FCC metals with low SFE along depth direction. The grain refinement mechanism along depth direction by multiple LSP impacts can be schematically illustrated in Fig. 8.6, in which each state will be discussed in terms of the experimental observations.

During multiple LSP impacts, dislocation activities, namely the pile up of DLs, lead to the formation of the SFs and PDAs with multiple directions in original grains, as can be clearly seen in Fig. 8.3e–g. With further increasing strain rate and strain, there are plenty of intersections of PDA-PDA (or PDA-SF), which gradually results in the subdivision of original coarse grains by forming submicron rhombic blocks (or submicron triangular blocks) primarily separated by PDAs or SFs, as can be clearly seen in Fig. 8.3e and f. As the SPD layer is deformed at a

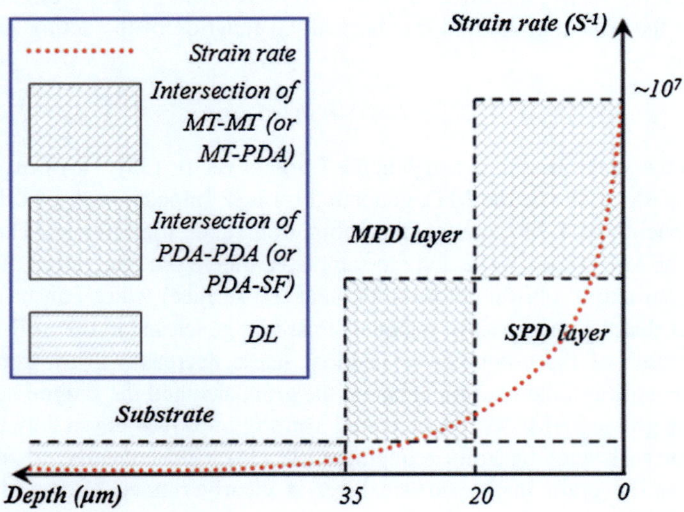


**Fig. 8.6** Schematic illustration showing micro-structural evolution process along depth direction induced by multiple LSP impacts

ultra-high strain rate, the intersection of MT-MT (or MT-PDA) is the common micro-structure, which subdivides the coarse grain into submicron rhombic blocks or submicron triangular blocks by the micro-MTs (or PDAs) with multiple directions, as can be clearly seen in Fig. 8.3a–d. With further increasing strain rate and strain, these MTs of submicron subdivided blocks turn gradually into subgrain boundaries, which eventually transform into equiaxed refined grains by the development of subgrain boundaries. The strain rate drops remarkably with an increase of depth from the top surface during multiple LSP impacts, and there are different micro-structure in different depths, so the strain and strain rate play an important role in the grain refinement process and the final stabilized grain size upon plastic deformation [19, 22].

Based on the micro-structure features observed in various layers with different strains and strain rates in the plastic deformation layer, the following elemental states are involved in the grain refinement process: (1) formation of PDAs and SFs with multiple directions due to the pile up of DLs; (2) intersections of PDA-SF (or PDA-PDA) with multiple directions; (3) formation of submicron triangular blocks (or irregularly shaped blocks) by the intersections of MT-MT (or MT-PDA) with multiple directions; (4) transformation of MTs into subgrain boundaries; (5) evolution of the continuous DRX in subgrain boundaries to refined grain boundaries.

It is well known that compressive residual stress can resist both crack initiation and small crack propagation, and the deeper surface layer with compressive residual stress by LSP technology contributes to the improvement of fatigue properties [5, 31]. In fact, the presence of compressive residual stresses is often the result of micro-plastic deformation accompanying the micro-structure changes [32]. As a result, it is reasonable to assume that the LSP induced strengthening in the FCC materials (AISI 304 SS) with very low SFE is due to the generation of MT-MT intersections with multiple directions.

The spacing of MTs is directly related to the refined structure dimensions ($L$) formed in the coarse grain, and it is basically a function of the acting shear stress ($\tau$), where

$$L = 10\,Gb/\tau \tag{8.1}$$

here $G$ is the shear modulus, and $b$ is the Burgers vector [33]. Obviously, with an increase of shear stress, the MTs generated by LSP impacts increases, leading to smaller spacing of T-M lamellae and submicron subdivided blocks. The reaction between the laser shock wave and the sample results in the generation of MTs and the micro-structural plastic deformation near the surface, which can be explained by the fact that the compressive residual stresses is generated in the SPD layer, and the magnitude of the compressive residual stress decreases away from the top surface. In addition, the relation between the grain size and the fatigue behavior of a ultra-fine grained AISI 304 SS indicated a strong improvement in both the tensile and fatigue resistance by grain refinement [32, 34]. Consequently, it can be concluded that the grain in the surface layer is clearly refined after a single LSP impact, which is favorable to the improvement on the fatigue life of AISI 304 SS.

It is interesting to mention that the refined structure at the top surface of the 304 SS is submicron triangular blocks after three LSP impacts, that is to say, MTs with three directions subdivide the coarse grain into submicron triangular blocks. If the above top surface of the 304 SS is treated by the fourth LSP impact (or the impact time >4), can the generated MTs in the fourth (nth) direction subdivide submicron triangular block into irregularly shaped submicron blocks with large misorientations? It is worth noting how to control the direction of MTs generated by each LSP impact. Further systematic investigation is needed to demonstrate these phenomenons.

## 8.8 Summary

In this chapter, the effects of multiple LSP impacts on the plastic deformation layer of AISI 304 SS are carried out, and the refined structure is obtained after multiple LSP impacts. Some important conclusions can be made as follows:

1. The intersectional micro-structure with multiple directions plays an important role in the grain refinement of coarse grains in the plastic deformation layer of the f.c.c. materials with low SFE along depth direction. The grain in the surface layer of AISI 304 SS subjected to multiple LSP impacts is extremely refined, and the grain size is about 50–200 nm at the top surface.
2. Multi-directional mechanical MT-MT intersections lead to grain subdivision at the top surface during multiple LSP impacts, and the novel structure with submicron triangular block is found at the top surface subjected to three LSP impacts.
3. A grain refinement mechanism induced by plastic deformation during multiple LSP impacts in AISI 304 SS is proposed based on the micro-structure observations. It involves the formation of PDAs and SFs with multiple directions due to the pile up of DLs, the formation of submicron triangular blocks by the intersections of MT-MT (or MT-PDA, or PDA-PDA) with multiple directions, transformation of MTs into subgrain boundaries, and evolution of the continuous DRX in subgrain boundaries to refined grain boundaries.
4. Experimental evidence and analysis of the grain refinement mechanism indicate that the ultra-high strain rate ultra-short laser pulse is a crucial factor in the formation of refined grains during plastic deformation of the 304 SS subjected to multiple LSP impacts.

## References

1. Lindemann, J., Buque, C., & Appel, F. (2006). Effect of shot peening on fatigue performance of a lamellar titanium aluminide alloy. *Acta Materialia, 54*(4), 1155–1164.
2. Bhattacharjee, P. P., Ray, R. K., & Tsuji, N. (2009). Cold rolling and recrystallization textures of a Ni–5 at % W alloy. *Acta Materialia, 57*(7), 2166–2179.

3. Venugopal, T., Rao, K. P., & Murty, B. S. (2007). Mechanical and electrical properties of Cu–Ta nanocomposites prepared by high-energy ball milling. *Acta Materialia, 55*(13), 4439–4445.
4. Lin, Y. M., Lu, J., Wang, L. P., Xu, T., & Xue, Q. J. (2006). Surface nanocrystallization by surface mechanical attrition treatment and its effect on structure and properties of plasma nitrided AISI 321 stainless steel. *Acta Materialia, 54*(20), 5599–5605.
5. Montross, C. S., Ye, L., Wei, T., Clark, G., & Mai, Y. W. (2002). Laser shock processing and its effects on microstructure and properties of metal alloys: A review. *International Journal of Fatigue, 24*, 1021–1036.
6. Meyers, M. A., Gregori, F., Kad, B. K., Schneider, M. S., Kalantar, D. H., Remington, B. A., et al. (2003). Laser-induced shock compression of monocrystalline copper: Characterization and analysis. *Acta Materialia, 51*(5), 1211–1228.
7. Zhang, H., & Yu, C. Y. (1998). Laser shock processing of 2024–T62 aluminum alloy. *Materials Science and Engineering A, 257*, 322–327.
8. Zhang, Y. K., Hu, C. L., Cai, L., Yang, J. C., & Zhang, X. R. (2001). Mechanism of improvement on fatigue life of metal by laser-excited shock waves. *Applied Physics A, 72*(2), 113–116.
9. Yilbas, B. S., Shuja, S. Z., Arif, A., & Gondal, M. A. (2003). Laser-shock processing of steel. *Journal of Materials Processing Technology, 135*(1), 6–17.
10. Srinivasan, S., Garcia, D. B., Gean, M. C., Murthy, H., & Farris, T. N. (2009). Fretting fatigue of laser shock peened Ti–6Al–4 V. *Tribology International, 42*(9), 1324–1329.
11. Nikitin, I., & Altenberger, I. (2007). Comparison of the fatigue behavior and residual stress stability of laser-shock peened and deep rolled austenitic stainless steel AISI 304 in the temperature range 25–600 °C. *Materials Science and Engineering A, 465*(1–2), 176–182.
12. Peyre, P., Scherpereel, X., Berthe, L., Carboni, C., Fabbro, R., Béranger, G., et al. (2000). Surface modifications induced in 316L steel by laser peening and shot-peening. Influence on pitting corrosion resistance. *Materials Science and Engineering A, 280*(2), 294–302.
13. Nikitin, I., Scholtes, B., Maier, H. J., & Altenberger, I. (2004). High temperature fatigue behavior and residual stress stability of laser-shock peened and deep rolled austenitic steel AISI 304. *Scripta Materialia, 50*(10), 1345–1350.
14. Sano, Y. J., Obata, M., Kubo, T., Mukai, N., Yoda, M., Masaki, K., et al. (2006). Retardation of crack initiation and growth in austenitic stainless steels by laser peening without protective coating. *Materials Science and Engineering A, 417*(1–2), 334–340.
15. Mordyuk, B. N., Milman, Y. V., Iefimov, M. O., Prokopenko, G. I., Silberschmidt, V. V., Danylenko, M. I., et al. (2008). Characterization of ultrasonically peened and laser-shock peened surface layers of AISI 321 stainless steel. *Surface and Coatings Technology, 202*(19), 4875–4883.
16. Masse, J. E., & Barreau, G. (1995). Surface modification by laser induced shock waves. *Surface Engineering, 11*, 131–142.
17. Ding, K., & Ye, L. (2003). Three-dimensional dynamic finite element analysis of multiple laser shock peening processes. *Surface Engineering, 19*(5), 351–358.
18. Chu, J. P., Rigsbee, J. M., Banas′, G., & Elsayed-Ali, H. E. (1999). Laser-shock processing effects on surface microstructure and mechanical properties of low carbon steel. *Materials Science and Engineering A, 260*, 260–268.
19. Lu, J. Z., Luo, K. Y., Zhang, Y. K., Cui, C. Y., Sun, G. F., Zhou, J. Z., et al. (2010). Grain refinement of LY2 aluminum alloy induced by ultra-high plastic strain during multiple laser shock processing impacts. *Acta Materialia, 58*(11), 3984–3994.
20. Tan, Y., Wu, G., Yang, J. M., & Pan, T. (2004). Laser shock peening on fatigue crack growth behavior of aluminum alloy. *Fatigue and Fracture of Engineering Materials and Structures, 27*(8), 649–656.
21. Arif, A. F. M. (2003). Numerical prediction of plastic deformation and residual stresses induced by laser shock processing. *Journal of Materials Processing Technology, 136*, 120–138.

22. Tao, N. R., Wang, Z. B., Tong, W. P., Sui, M. L., Lu, J., & Lu, K. (2002). An investigation of surface nanocrystallization mechanism in Fe induced by surface mechanical attrition treatment. *Acta Materialia, 50*(18), 4603–4616.
23. Sun, H. Q., Shi, Y. N., Zhang, M. X., & Lu, K. (2007). Plastic strain-induced grain refinement in the nanometer scale in a Mg alloy. *Scripta Mateialia, 55*, 975–982.
24. Wen, M., Liu, G., Gu, J. F., Guan, W. M., & Lu, J. (2009). Dislocation evolution in titanium during surface severe plastic deformation. *Applied Surface Science, 255*(12), 6097–6102.
25. Tao, N. R., & Lu, K. (2009). Nanoscale structural refinement via deformation twinning in face-centered cubic metals. *Scripta Materialia, 60*(12), 1039–1043.
26. Wu, X., Tao, N., Hong, Y., Liu, G., Xu, B., Lu, J., et al. (2005). Strain-induced grain refinement of cobalt during surface mechanical attrition treatment. *Acta Materialia, 53*(3), 681–691.
27. Zhang, H. W., Hei, Z. K., Liu, G., Lu, J., & Lu, K. (2003). Formation of nanostructured surface layer on AISI 304 stainless steel by means of surface mechanical attrition treatment. *Acta Materialia, 51*(7), 1871–1881.
28. Eddahbi, M., Del Valle, J. A., P'erez-Prado, M. T., & Ruano, O. A. (2005). Comparison of the microstructure and thermal stability of an AZ31 alloy processed by ECAP and large strain hot rolling. *Materials Science and Engineering A, 410–411*, 308–311.
29. Wang, Y. B., Louie, M., Cao, Y., Liao, X. Z., Li, H. J., Ringer, S. P., et al. (2010). High-pressure torsion induced microstructural evolution in a hexagonal close-packed Zr alloy. *Scripta Materialia, 62*(4), 214–217.
30. Wang, K., Tao, N. R., Liu, G., Lu, J., & Lu, K. (2006). Plastic strain-induced grain refinement at the nanometer scale in copper. *Acta Materialia, 54*(16), 5281–5291.
31. Zhang, X. C., Zhang, Y. K., Lu, J. Z., Xuan, F. Z., Wang, Z. D., & Tu, S. D. (2010). Improvement of fatigue life of Ti-6Al-4 V alloy by laser shock peening. *Materials Science and Engineering A, 527*(15), 3411–3415.
32. Belyakov, A., Tsuzaki, K., Miura, H., & Sakai, T. (2003). Effect of initial microstructures on grain refinement in a stainless steel by large strain deformation. *Acta Materialia, 51*(3), 847–861.
33. Kuhlmann-Wilsdorf, D., & Van der Merwe, J. H. (1982). Theory of dislocation cell size in deformed metals. *Materials Science and Engineering, 55*, 79–83.
34. Schino, A. D., & Kenny, J. M. (2003). Grain size dependence of the fatigue behavior of a ultrafine-grained AISI 304 stainless steel. *Materials Letters, 57*(21), 3182–3185.

# Chapter 9
# Electrochemical Corrosion Resistance of AISI 304 SS Weldment Treated by Laser Shock Processing

**Abstract** This chapter figures out effects of LSP on electrochemical corrosion resistance of AISI 304 SS weldments after cavitation erosion. Through the electrochemical test, AISI 304 SS weldments without and with LSP impacts after cavitation erosion are studied. Influence mechanism of LSP on electrochemical corrosion resistance of AISI 304 SS weldments after cavitation erosion is also revealed.

## 9.1 Introduction

Liquefied natural gas (LNG) is considered as superior, efficient, economical and clean energy. Since the late 1980s, LNG has been researched and applied into alternative fuel vehicles. Some companies, such as CHART-MVE Company (USA), have developed the skid-mounted package LNG station [1–4]. During working of the skid-mounted package LNG station, cavitation erosion can be found on the surface of cryopump for LNG. The process of cavitation erosion is complicated; in general, cavitation erosion is closely related to water hammer effect. As a result, it can be concluded that cavitation erosion is formed under the interaction of mechanical, chemical and electrochemical functions during water hammer effect [5–7]. Among of above-mentioned functions, electrochemical function plays an indispensable role. The relative significance and the interaction between erosion and corrosion may be synergistic in the case of metallic materials under the specific environment [8, 9]. In addition, due to reasonable mechanical properties and cheap price, AISI 304 SS is widely used for LNG utilization [10, 11]. Cavitation erosion occurs on the cryopump made of AISI 304 SS for LNG, especially for its weldments. Thus, electrochemical corrosion resistance of AISI 304 SS weldments after cavitation erosion should be improved.

LSP is one of the most important surface modification techniques and LSP is regarded as a competitively alternative technology [12, 13]. The high strength impact waves can be generated due to the interaction between laser beams (high

Y. Zhang et al., *Laser Shock Processing of FCC Metals*,
Springer Series in Materials Science 179, DOI: 10.1007/978-3-642-35674-2_9,

power density: GW/cm$^2$, short pulse: ns) and the surface of metallic materials. Thus, compressive residual stresses of several hundred MPa are induced [14]. Mechanical properties of metallic materials, such as fatigue [15], corrosion resistance [14, 16], wear [17] and etc., are improved by LSP impacts.

In this chapter, it is mainly trying to figure out effects of LSP on electrochemical corrosion resistance of AISI 304 SS weldments after cavitation erosion. Through the electrochemical test, AISI 304 SS weldments without and with LSP impacts after cavitation erosion are studied. These results are explained by the analysis of mechanical properties and microstructures.

## 9.2 Experimental Procedures

### *9.2.1 Experimental Material and Sample Preparation*

In this study, AISI 304 SS weldments with a thickness of 5 mm were experimental materials and obtained by using YLS-4000 fiber laser (IPG, USA). During laser welding process, appropriate laser welding parameters were chosen and given in Table 9.1. Welding joints of weldments were treated with full penetration, the average width of laser welding zone (WZ) at the frontage of weldments was 1.12 mm, and that at the back was 1.20 mm. The width of heat-affected zone (HAZ) was small and the laser welding surface was smooth without spatter. The schematic diagram of laser weldment is shown in Fig. 9.1.

### *9.2.2 Experimental Procedure of Laser Welding and LSP*

During the whole experiment, the position of the WZ, HAZ and base metal (BM) for weldments was marked by marking pen carefully, in order to distinguish different zones of weldments well. And then, the surface of weldments were polished with SiC paper with different grades of roughness (from 150 to 1600), followed by cleaning in deionized water. Subsequently, the surface was degreased in ethanol by using ultrasonic cleaning. LSP experiments were carried out shortly after preparation.

Weldments were treated by a Q-switched Nd: YAG laser (GAIA-R, THALES) with a wavelength of 1064 nm and pulse duration of about 15 ns. The LSP process and the schematic diagrams were detailed clearly in literatures [12, 15, 18, 19].

**Table 9.1** Laser welding parameters chosen in the present study

| Laser type | Laser power P (kW) | Welding speed v (mm•s$^{-1}$) | Defocusing distance Δf (mm) |
|---|---|---|---|
| Fiber laser | 4 | 35 | □–3 |

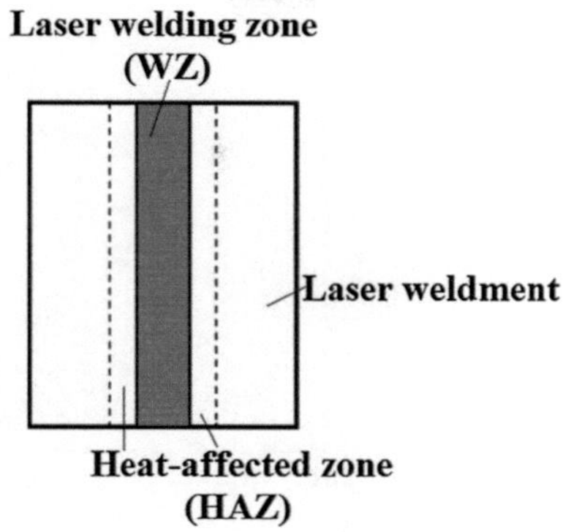


**Fig. 9.1** Schematic diagram of laser weldment

In this study, laser pulse energy is 6 J, laser spot diameter is 3 mm, and repetition-rate is 1 Hz. The polished surface of weldments was treated by LSP impacts, and the overlapping rate of the laser spot was 50 %.

In order to improve the absorption from the laser energy and prevent the weldment surface from the laser ablation, the aluminum foil with a thickness of 0.1 mm was jointed on the weldment surface closely as an absorbing layer. At the same time, the water layer with a thickness of 1–2 mm was used as the transparent confining layer to increase the peak pressure of laser shock waves, impulse in samples and the action time [20–22].

### *9.2.3 Cavitation Erosion Testing and Residual Stress Measurements*

Weldments without and with LSP impacts were treated by ultrasonic vibratory tests using a tip of a vibrating horn submerged in distilled water at the room temperature. Cavitation erosion testing parameters were presented in Table 9.2. The distance between weldments and the tip of the vibrating horn was 10 mm. The cavitation erosion testing lasted for at least 6 h. The process sequence graph (PSG) of the above-mentioned LSP and cavitation erosion process was given in Fig. 9.2.

Measurements of residual stress depth profile in the WZ and HAZ of weldments without and with LSP impacts were measured using X-ray diffraction ($\sin^2\psi$ method). The X-ray beam diameter was about 1 mm. X-ray source was Cr-$K_\beta$ ray and the diffraction plane was $\beta$ phase (311) plane. In the stress calculation, the Poisson's ratio was set to be 0.35. The speed of the ladder scanning was 0.1 %/s. The scanning starting angle and terminating angle were about 145 and 153°,

**Table 9.2** Experimental parameters in cavitation erosion testing

| Diameter of vibrating horn (mm) | Frequency (kHz) | Temperature (°C) | Total testing time (h) |
|---|---|---|---|
| 20 | 19–21 | 20 | At least 6 |

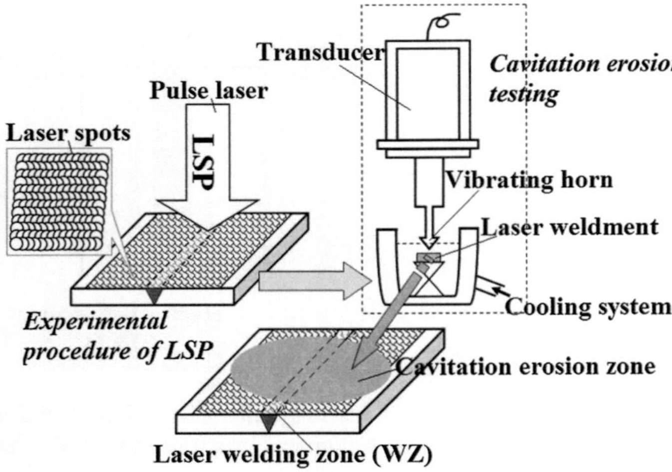


**Fig. 9.2** The PSG of LSP and cavitation erosion

respectively. During residual stress depth profile measurements, the surface of weldments was removed by electrolytic polishing to obtain surface residual stresses from the mechanical polishing. Residual stresses determined using XRD were arithmetic average stresses in a volume of material defined by the irradiated area and the depth of penetration of the x-ray beam. Measurements were repeated three times for each condition, and average values were obtained.

### 9.2.4 Morphology Observation and Surface Roughness Measurements

Micrographs of cross sections in the WZ and HAZ of weldments without and with LSP impacts after cavitation erosion were observed by using scanning electron microscope (SEM) (JSM-7001F, JEOL, Japan).

The surface roughness in the WZ and HAZ of weldments without and with LSP impacts after cavitation erosion was measured by roughness tester (surfcom 130A-Monochrome, KA1314MF). Figure 9.3 shows measurement direction of the surface roughness.

### 9.2.5 Electrochemical Testing and Surface Morphology Observation

The wire-electrode cutting was used to cut the above-mentioned processed weldments without and with LSP impacts after cavitation erosion into small pieces with the dimension of 10 × 10 mm along the direction parallel to the WZ. These

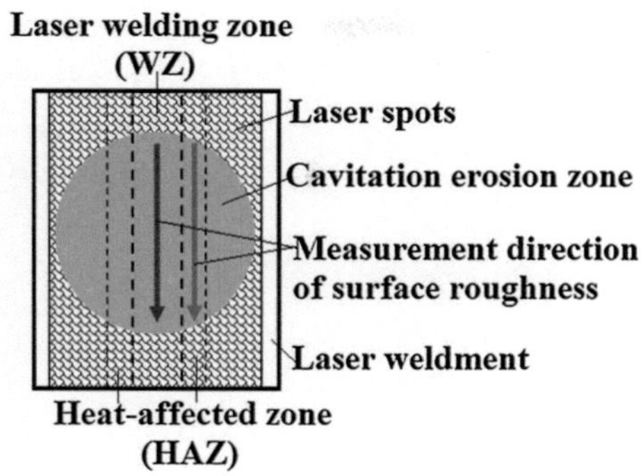


**Fig. 9.3** Measurement direction of surface roughness by using roughness tester

small pieces were separated into two groups of samples, the first group of samples only contained WZ in the middle of weldments, and the second group of samples contained HAZ alongside WZ and BM, as shown in Fig. 9.4. The preparation of samples was convenient for the next electrochemical testing, which were prepared as the working electrode.

Electrochemical testing was carried out to investigate the electrochemical corrosion behavior of weldments without and with LSP impacts after cavitation erosion in 3.5 wt % NaCl solution, as shown in Fig. 9.5. Potentiodynamic polarization scans were carried out using an electrochemical system according to the ASTM Standard G5-94 [23]. During electrochemical testing, the solution was maintained at room temperature. The reference electrode was a saturated calomel electrode (SCE) with luggin capillary, and platinum foil (Pt) was used as a counter electrode for current measurement. The above-mentioned two groups of samples, with only containing WZ or HAZ, were packed by polyethylene ring full-filled with AB rubber. The working face with the dimension of 10 $\times$ 10 mm and copper wire were left, which were prepared for working electrode during the electrochemical testing, respectively. A constant potential scan rate of 0.5 mV/s was used for all the potentiodynamic polarization testing. All data were recorded after an initial delay time of 10 min for weldments to reach a steady state. Corrosion current densities were determined from polarization curves using the Tafel extrapolation method. All the electrochemical testing was repeated to obtain reproducibility of results.

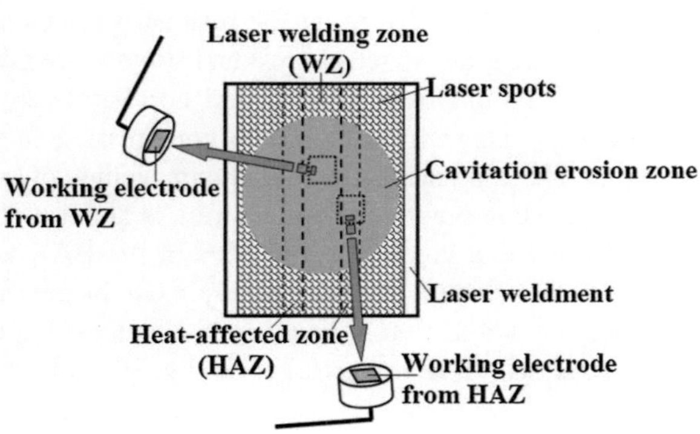


**Fig. 9.4** Preparation of working electrode for electrochemical testing

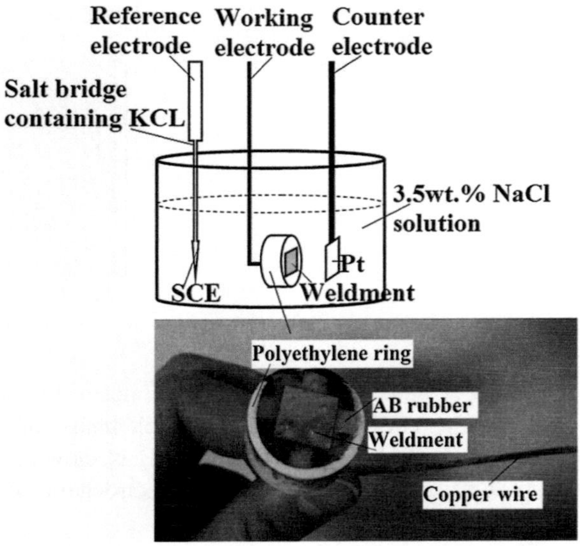


**Fig. 9.5** Schematic diagram of the electrochemical testing system

Through the above-mentioned electrochemical testing, surface morphologies of weldments without and with LSP impacts after cavitation erosion were characterized by using optical microscope (OM) (DM 2500 M, LEICA).

## 9.3 Results and Discussion

### *9.3.1 Residual Stress Depth Profile of Weldments Without and with LSP Impacts*

Figure 9.6 shows residual stress depth profile of weldments without and with LSP impacts. It can be seen from Fig. 9.6 that residual stresses in the WZ and HAZ without LSP impacts are approximately in a state of tensile residual stress. But the surface tensile residual stress in the WZ is smaller than that in the HAZ. Surface failure often occurs due to tensile residual stresses developed on the outside surface of the material, so tensile residual stresses are generally deemed to deteriorate the fatigue failure from erosion and corrosion [24]. However, residual stresses remarkably change from tensile to compressive in the WZ and HAZ with LSP impacts. The top surface has maximum values of compressive residual stresses. The surface compressive residual stress is about −400 MPa in the WZ with LSP impacts, and that is about −340 MPa in the HAZ with LSP impacts. Significant compressive residual stresses mainly exist in near-surface regions for all cases [25]. On the whole, compressive residual stresses in the WZ are higher than those in the HAZ with LSP impacts. In addition, it can be concluded that the depth of the

Fig. 9.6 Residual stress depth profile of weldments without and with LSP impacts

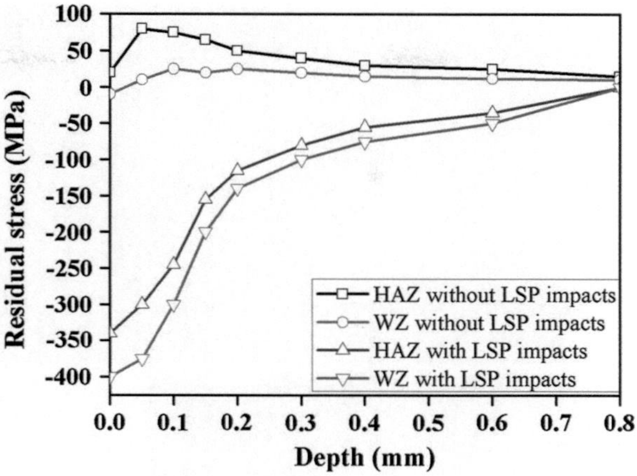


compressive residual stress gets up to about 0.8 mm from the surface. Thus, high-level compressive residual stresses can be induced by laser shock waves, and LSP impacts can effectively convert tensile residual stresses to relatively strong compressive residual stresses. Compressive residual stresses can delay crack initiation on the surface, and slow the growth of micro-crack in the depth direction [26, 27]. High-level compressive residual stresses induced by LSP impacts are favorable to improve the erosion and corrosion resistance of weldments. It is well known that compressive residual stresses are generated due to the local plastic deformation with LSP impacts [28].

### *9.3.2 Morphology Observation of Cross Sections of Weldments Without and with LSP Impacts After Cavitation Erosion*

Figure 9.7 shows SEM morphology observation of cross sections in the WZ without and with LSP impacts after cavitation erosion. The most obvious change of the material microstructures is in near-surface regions [21], moreover, the affected depth of the material by cavitation erosion is much smaller than that by LSP impacts [29]. Thus, the change of the material microstructure is primarily induced by LSP impacts in near-surface regions. Figure 9.7a shows SEM morphology observation of cross sections in the WZ without LSP impacts after cavitation erosion in near-surface regions. Figure 9.7b shows SEM morphology observation of cross sections in the WZ with LSP impacts after cavitation erosion in near-surface regions, and Fig. 9.7c is the high magnification SEM micrograph of Fig. 9.7b. From Fig. 9.7a–c, it can be found that the columnar austenite is formed due to the first crystallization of the austenite from the liquid state to the solid state in the WZ. And then, the secondary crystallization appears when the

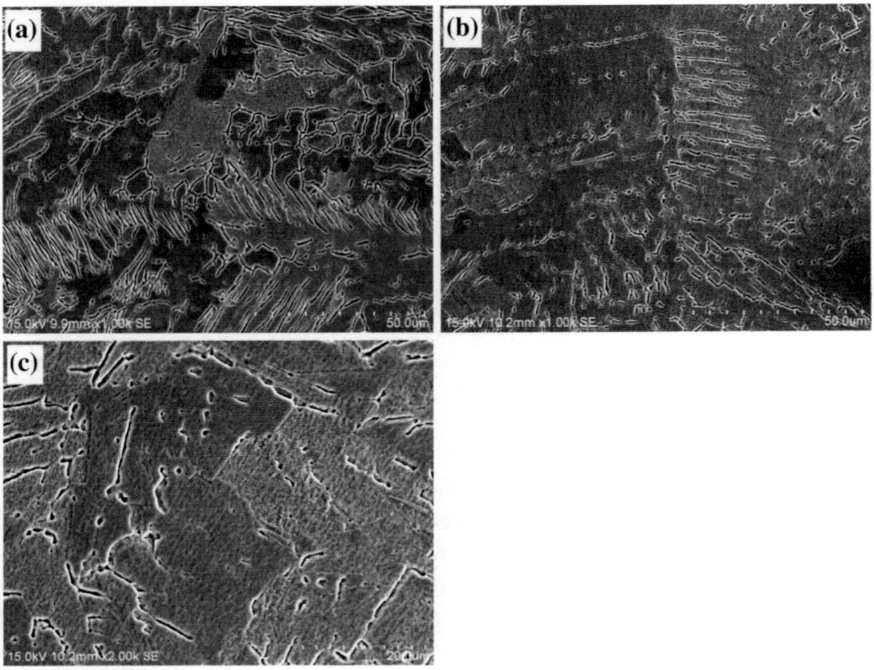


**Fig. 9.7** SEM morphology observation of cross sections in the WZ after cavitation erosion **a** without LSP impacts, **b** with LSP impacts, and **c** the high magnification SEM micrograph of Fig. 9.7b

temperature is cooled to the room temperature in the WZ. Thus, some microstructures, such as the ferrite, will be separated from the above-mentioned columnar austenite during the cooling process. The eutectoid ferrite is precipitated along austenite grain boundaries, sequentially the columnar grain profile of the original austenite is delineated, and the ferrite is the columnar ferrite [30, 31]. Grains of the columnar ferrite and the original austenite without LSP impacts are coarse, as shown in Fig. 9.7a. While grains of the columnar ferrite and the original austenite with LSP impacts are refined, as shown in Fig. 9.7b. The fine and dense ferrite is best hoped to appear in the WZ, which is beneficial to enhance mechanical properties of weldments, especially for the improvement of the impact toughness [32, 33]. Above all, large quantities of slip systems occur in the original austenite, as shown in Fig. 9.7c. The laser strengthening layer is generated in the surface and near-surface regions of weldments with LSP impacts, that is, the compressive residual stress layer, which plays a decisive role in improving the erosion and corrosion of weldments.

Figure 9.8a and b show SEM morphology observation of cross sections in the HAZ without and with LSP impacts after cavitation erosion, respectively. Only the coexistence between the columnar ferrite and the austenite exists in the HAZ of weldments without LSP impacts, as shown in Fig. 9.8a. While grains are also refined in the HAZ with LSP impacts; most of all, lots of slip systems occur in

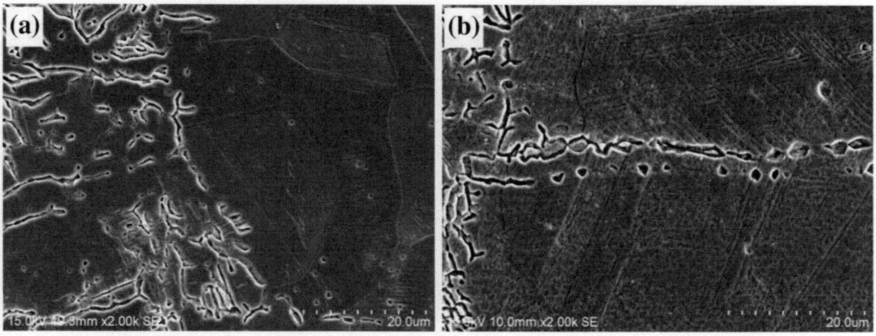


**Fig. 9.8** SEM morphology observation of cross sections in the HAZ after cavitation erosion **a** without LSP impacts, and **b** with LSP impacts

grains, as shown in Fig. 9.8b. Thus, the principle of the plastic deformation induced by LSP impacts in the HAZ is similar to that in the WZ. On the one hand, high-level compressive residual stresses are induced by LSP impacts (as shown in Sect. 9.3.1), so the relative movement of grain boundaries is held back effectively, and the intergranular deformation becomes difficult. Compressive residual stresses induced by LSP impacts also can completely compensate for or widely decrease tensile additive stresses generated by the nonuniform deformation, which relieves undesirable effects of tensile stresses and makes the material become denser [14]. On the other hand, LSP owns the ultra-high strain rate, and extremely shorten the time difference about grains from the slip deformation to the plastic deformation in laser weldments. Therefore, not only are disadvantages of the deformation nonuniformity eliminated to a greater extent, but also the ultra-high strain rate ($10^7$ $s^{-1}$) of LSP increases the dislocation density [21]. The dislocation movement makes the plastic deformation occur in the material through the slip. These dislocations are activated in grains and move along the slip direction in the slip plane. When dislocations disappear, steps are formed. Generally speaking, dislocations move to grain boundaries and stop, which induces the lattice distortion, and new grain boundaries are formed. During this process, more slips are needed to cooperate with the dislocation movement, and then grains are refined [21, 34, 35]. From Fig. 9.7 in the WZ and Fig. 9.8 in the HAZ, it can be found that for the plastic deformation of laser weldments many micro steps are generated due to the slip. Through LSP impacts, slip systems become more and denser, which also can improve the plasticity of laser weldments.

### 9.3.3 Comparisons Between Surface Roughness of Weldments Without and with LSP Impacts After Cavitation Erosion

Figure 9.9a and b show surface roughness of weldments without and with LSP impacts after cavitation erosion. Figure 9.9a shows surface roughness in the WZ without and with LSP impacts after cavitation erosion. It can be seen that after

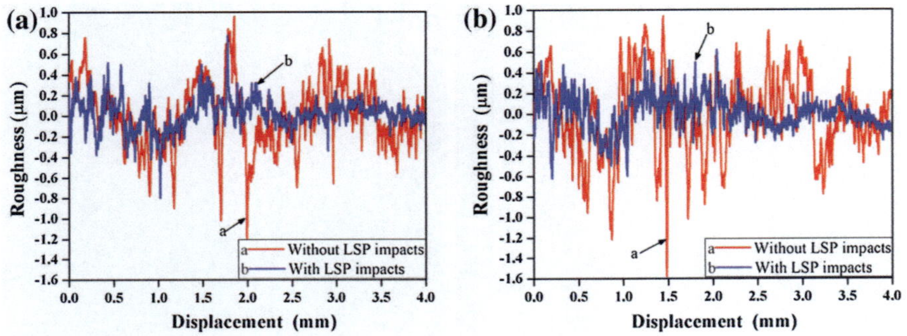


**Fig. 9.9** Surface roughness of weldments without and with LSP impacts after cavitation erosion **a** in the WZ, and **b** in the HAZ

cavitation erosion the surface roughness in the WZ with LSP impacts is smaller than that without impacts. Figure 9.9b shows surface roughness in the HAZ without and with LSP impacts after cavitation erosion. The surface roughness in the HAZ with LSP impacts is also smaller than that without impacts after cavitation erosion. In addition, on the basis of comparison between Fig. 9.9a and b, it can be found that the surface roughness in the WZ is smaller than that in the HAZ after cavitation erosion as a whole. It is well known that local stress concentrations will be generated because of the higher surface roughness, which leads to the initiation and growth of fatigue cracks [36]. HAZ with higher surface roughness is more prone to be damaged by cavitation erosion [37, 38]. As a result, cavitation erosion resistance in the WZ is better than that in the HAZ [39]. Above all, during LSP it generates high strength impact waves and induces compressive residual stresses of several hundred MPa (as shown in Sect. 9.3.1) by high power density ($GW/cm^2$), short pulse (ns level) laser beams [14]. Thus, the surface roughness with LSP impacts is smaller than that without LSP impacts, thus the surface roughness with LSP impacts decreases, which may improve fatigue lives of the material [37].

### *9.3.4 Analysis of Potentiodynamic Polarization Curves for Laser Weldments Without and with LSP Impacts After Cavitation Erosion*

Figure 9.10 shows potentiodynamic polarization curves in the WZ and HAZ of weldments without and with LSP impacts after cavitation erosion in 3.5 wt % NaCl solution, respectively. Anodic and cathodic portions of potentiodynamic polarization curves are obtained when the free-corrosion potential is swept at first. Curves (a) and (b) indicate potentiodynamic polarization curves in the HAZ and WZ of weldments without LSP impacts after cavitation erosion, respectively. Curves (c) and (d) indicate potentiodynamic polarization curves in the HAZ and

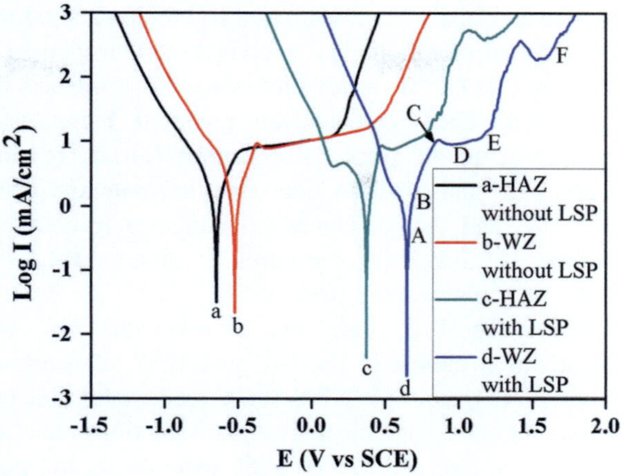


**Fig. 9.10** Polarization curves in the WZ and HAZ of AISI 304 SS weldments without and with LSP impacts after cavitation erosion in 3.5 wt % NaCl solution, respectively

WZ of weldments with LSP impacts after cavitation erosion, respectively. From Fig. 9.10, it can be found that LSP impacts play a large role on electrochemical corrosion resistance of AISI 304 SS weldments after cavitation erosion. In addition, the free-corrosion current can be figured out by Tafel extrapolation method, and values of the free-corrosion potential and free-corrosion current are given in Table 9.3.

Based on the combination of Fig. 9.10 and Table 9.3, properties of corrosion resistance in the WZ are better than those in the HAZ without LSP impacts after cavitation erosion, and it can be seen that anodic portion of Curve (b) is on the right below Curve (a) in Fig. 9.10. The free-corrosion potential of Curve (a) is about −648 mV, and that of Curve (b) is about −522 mV. The free-corrosion current of Curve (a) is about 31.6 μA/cm$^2$, and that of Curve (b) is about 20.0 μA/cm$^2$. This result also proves that the HAZ is more vulnerable to erosion and corrosion because of high tensile residual stresses (as shown in Sect. 9.3.1). In most instances, erosion and corrosion phenomena are generated in areas where surface residual stresses are high residual stresses. Boven et al. [40] found the peak occurrence of pitting was located at the highest tensile residual stress. Turnbull et al. [41] indicated that the propensity for cracking also could be linked to the high residual stress at the surface. In other words, the formation of micro-pitting occurs preferentially in areas where tensile residual stresses are the highest. With respect to areas of lower residual stress levels, areas of higher residual stresses are

**Table 9.3** Corrosion properties in the WZ and HAZ of AISI 304 SS weldments without and with LSP impacts after cavitation erosion in 3.5 wt % NaCl solution

| | a | b | c | d |
|---|---|---|---|---|
| | Without LSP impacts | | With LSP impacts | |
| | Heat-affected zone (HAZ) | Laser welding zone (WZ) | Heat-affected zone (HAZ) | Laser welding zone (WZ) |
| $E_{corr}$(mV) | −648 | −522 | 372 | 646 |
| $I_{corr}$(μA/cm$^2$) | 31.6 | 20.0 | 4.0 | 2.5 |

anodic and easily to be destroyed by erosion and corrosion, such as the appearance of corroded morphologies. In comparison, properties of corrosion resistance in the WZ and HAZ of weldments with LSP impacts after cavitation erosion are improved. The free-corrosion potential increases and free-corrosion current decreases, anodic portions of Curves (c) and (d) move to the right side below Curves (a) and (b). The free-corrosion potential of Curve (c) increases to about 372 mV, and that of Curve (d) increases to about 646 mV. The free-corrosion current of Curve (c) decreases to about 4.0 $\mu A/cm^2$, and that of Curve (d) decreases to about 2.5 $\mu A/cm^2$.

From Fig. 9.10, it also can be found that obvious passivation areas occur on polarization curves in the WZ and HAZ of weldments with LSP impacts after cavitation erosion, including the characteristic of activation, passivation and over-passivation [42], and the activated dissolution area is short. Curve (d) is put forward as an example, and the cathodic reaction on the polarization curve is hydrogen evolution reaction (HER) for the free-corrosion potential in 3.5 wt % NaCl solution, while the anodic reaction is dissolution reaction of weldments. As for Curve (d), Portion AC is the activated dissolution area. AISI 304 SS weldments are considered as anode lose electron, because the potential of working electrode is higher than that of weldments in such circumstances of thermodynamics equilibrium. Thus, the oxidation reaction is generated, which is also known as electrochemical dissolution reaction [43]. The anodic current density increases with the increment of the potential. Thereinto, Portion AB is the linear Tafel region [44]. After Point B, the slope of the polarization curve gradually decreases, and that reaches zero at Point C. It can be said that the growth rate of the current density becomes slow. The potential is 874 mV, the anodic current density reaches the maximum value on Portion AD. Portion CD is the transitional area between activation area and passivation area, the anodic current density decreases gradually with the increment of the potential. The anodic dissolution rate is restricted by the generation and dissolution reaction of the passive film on the surface of weldments, and this restriction is enhanced as the potential increases. Portion DE is the passivation area. The anodic current density stays the same as the potential increases, which indicates that there is a dynamic balance between the generation and dissolution reaction of the passive film [45]. Portion EF is over-passivation area, the anodic current density increases quickly with the increment of the potential. When the potential reaches 1400 mV, the anodic current density exhibits an oscillation phenomenon [46].

### *9.3.5 OM Micrographs on the Surface in the WZ and HAZ of Laser Weldments Without and with LSP Impacts After Cavitation Erosion in 3.5 wt % NaCl Solution*

Figure 9.11a–d show corroded morphologies in the WZ and HAZ of AISI 304 SS weldments without and with LSP impacts after cavitation erosion in 3.5 wt % NaCl solution, respectively. Figure 9.11a shows corroded morphologies in the WZ

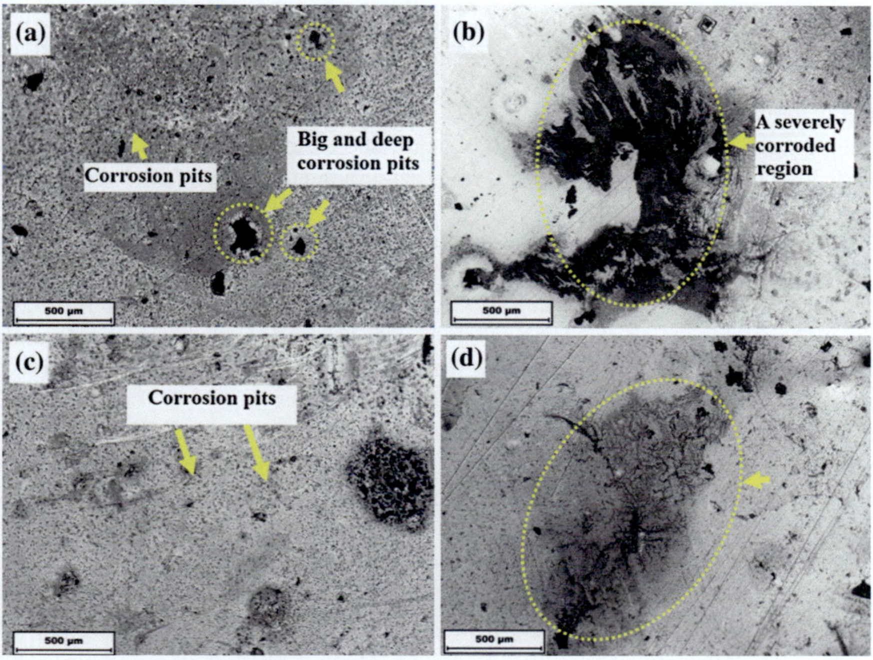


**Fig. 9.11** Corroded morphologies in the WZ and HAZ of AISI 304 SS weldments without and with LSP impacts after cavitation erosion in 3.5 wt % NaCl solution, respectively **a** in the WZ without LSP impacts, **b** in the HAZ without LSP impacts, **c** in the WZ with LSP impacts, and **d** in the HAZ with LSP impacts

without LSP impacts after cavitation erosion in 3.5 wt % NaCl solution. A lot of big and deep corrosion pits appear. During laser welding processing, the coarse ferrite increases with microsegregation of chromium at ferrite-austenite interphase boundaries, which causes chromium-depleted zones formed [47, 48]. In addition, the surface in the WZ is further corroded by the water hammer effect during cavitation erosion processing and the chloride ions from 3.5 wt % NaCl solution. Thus, a large number of corrosion pits in the WZ are generated. Figure 9.11c shows corroded morphologies in the WZ with LSP impacts after cavitation erosion in 3.5 wt % NaCl solution. By comparison, mechanical properties of the surface in the WZ are strengthened through LSP impacts, such as the increment of the hardness [49] and compressive residual stresses (as shown in Sect. 9.3.1), the decrease of the surface roughness (as shown in Sect. 9.3.3), and the refinement of the coarse ferrite (as shown in Sect. 9.3.2). Thus, surface morphologies in Fig. 9.11c are different from those in Fig. 9.11a. The distribution and size of corrosion pits are uniform, and the size of corrosion pits is smaller than that without LSP impacts. The free-corrosion potential involved in Sect. 9.3.4 is related to the formation of corrosion pits on the surface in the WZ. Due to the formation of new corrosion pits and the spread of corrosion products, values of the free-corrosion potential are to more negative. Moreover, corrosion pits grow seriously as the current density is higher [50–53]. According to the relationship between corroded morphologies and electrochemical properties, the free-corrosion potential in Fig. 9.11c is higher than that in Fig. 9.11a and the free-corrosion current in Fig. 9.11c is lower than that in Fig. 9.11a after the analysis of potentiodynamic polarization curves in Sect. 3.4. In a word, the electrochemical damage of the surface in the WZ with LSP impacts after cavitation erosion is smaller than that in the WZ without LSP impacts.

Figure 9.11b shows corroded morphologies in the HAZ without LSP impacts after cavitation erosion in 3.5 wt % NaCl solution. Corroded morphologies in the HAZ are different from those in the WZ. The multicorrosion process appearing in the WZ is not seen in the HAZ. From Fig. 9.11a and c, there are some single corrosion pits one by one in the WZ, while there is a severely corroded region in the HAZ, as shown in Fig. 9.11b. Due to effects of cavitation erosion and electrochemical corrosion, corrosion products spread out continuously around the initial damage at firstly, and subsequently, a combination of the above-mentioned erosion and corrosion affects a large area of the surface in the HAZ. Combined with the data in Sect. 3.4, the sharp shift of the potential from positive values to negative ones indicates the formation of pits on the surface in the HAZ [54]. If the potential increases, a piece of corroded region in the HAZ neither repassivates nor propagates as active (etching) pits [55]. Figure 9.11d shows corroded morphologies in the HAZ with LSP impacts after cavitation erosion in 3.5 wt % NaCl solution. The severely corroded region becomes smaller and shallower than that in Fig. 9.11b. By comparison, it can be gained that LSP impacts can improve the electrochemical corrosion resistance of AISI 304 SS weldments after cavitation erosion.

Figure 9.12 shows high magnification OM micrographs of the Fig. 9.11. It is found clearly that pitting corrosion occurs in the WZ of AISI 304 SS weldments without and with LSP impacts after cavitation erosion in 3.5 wt % NaCl solution [56], respectively, as shown in Fig. 9.12a and c. Specifically, the size and quantity of corrosion pits in Fig. 9.12a are more than those in Fig. 9.12c. In addition, the distribution of corrosion pits in Fig. 9.12a is disordered, while that in Fig. 9.12c is homogeneous, which is in favor of restraining the quick formation of corrosion pits on the surface in the WZ [57]. Corrosion pits in the HAZ are formed around the original initiation site and grow by a layered or oscillating type of process in Fig. 9.12b. The single initiation pit is surrounded by a lot of big corrosion pits and these corrosion pits grow in NaCl solution with a tendency for repassivation to appear. Thus, the surface in the HAZ without LSP impacts may collapse or become insufficient to provide the necessary diffusion barrier. The surface in the HAZ with LSP impacts is shown in Fig. 9.12d, the big severely corroded region is restrained to form due to LSP impacts, with some big corrosion pits left. It is considered that the surface in the HAZ with LSP impacts is strongly hindered from passing high anodic current densities. As a result, the electrochemical corrosion resistance of AISI 304 SS weldments after cavitation erosion is improved by LSP impacts.

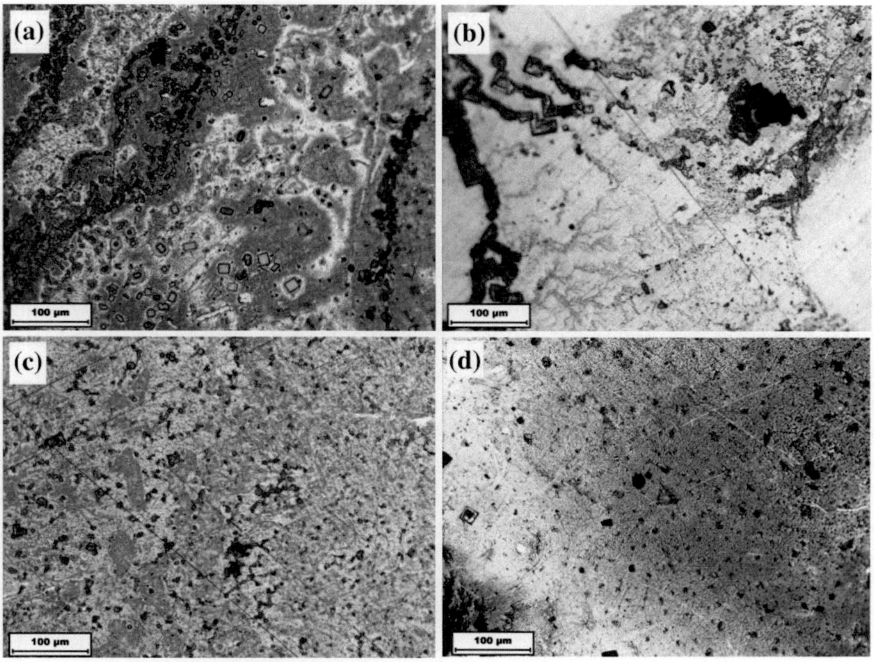


**Fig. 9.12** High magnification OM micrographs of Fig. 9.11a–d, respectively

## 9.4 Summary

This chapter presents the effects of LSP on electrochemical corrosion resistance of AISI 304 SS weldments after cavitation erosion. Electrochemical testing is carried out to investigate the electrochemical corrosion behavior of AISI 304 SS weldments without and with LSP impacts after cavitation erosion in 3.5 wt % NaCl solution. Some important conclusions can be made as follows:

(1) LSP impacts remarkably change residual stresses from tensile to compressive in the WZ and HAZ of weldments. The top surface has maximum values of compressive residual stresses. On the whole, compressive residual stresses in the WZ are higher than those in the HAZ with LSP impacts.

(2) Grains of the columnar ferrite and the original austenite with LSP impacts after cavitation erosion are refined in the WZ. For the plastic deformation in the WZ and HAZ with LSP impacts after cavitation erosion, many micro steps are generated due to the slip. Through LSP impacts, slip systems become more and denser.

(3) The surface roughness in the WZ and HAZ with LSP impacts is smaller than that without LSP impacts after cavitation erosion. The surface roughness in the WZ is smaller than that in the HAZ after cavitation erosion as a whole.

(4) The free-corrosion potential of potentiodynamic polarization curves increases from −648 mV in the HAZ without LSP impacts to 372 mV with LSP

impacts after cavitation erosion, and the free-corrosion potential increases from −522 mV in the WZ without LSP impacts to 646 mV with LSP impacts after cavitation erosion. The free-corrosion current of 31.6 μA/cm$^2$ in the HAZ without LSP impacts decreases to 4.0 μA/cm$^2$ with LSP impacts after cavitation erosion, and the free-corrosion current of 20.0 μA/cm$^2$ in the WZ without LSP impacts decreases to 2.5 μA/cm$^2$ with LSP impacts after cavitation erosion. After electrochemical test, obvious passivation areas occurre on polarization curves in the WZ and HAZ of weldments with LSP impacts after cavitation erosion.

(5) Pitting corrosion occurs in the WZ of AISI 304 SS weldments without and with LSP impacts after cavitation erosion in 3.5 wt % NaCl solution. Compared with corroded morphologies in the WZ without LSP impacts after cavitation erosion in 3.5 wt % NaCl solution, the distribution and size of corrosion pits are uniform, and the size of corrosion pits is smaller in the WZ with LSP impacts. Corroded morphologies in the HAZ are different from those in the WZ. The multi-corrosion process appearing in the WZ is not seen in the HAZ. There is a severely corroded region in the HAZ without LSP impacts after cavitation erosion in 3.5 wt % NaCl solution. The severely corroded region becomes smaller and shallower, with some big corrosion pits left in the HAZ with LSP impacts.

## References


1. Li, J., Li, Z. H., & Hua, B. (2005). Current status and developing prospect of LNG cryogenic energy utilization. *Indian Natural Gas, 25*, 103–105.
2. Lai, Y. K. (2005). Developing prospect of LNG vehicles. *Indian Natural Gas, 25*, 104–106.
3. Shi, G. H., Jing, Y. Y., Wang, S. L., & Zhang, X. T. (2010). Development status of liquefied natural gas industry in China. *Energy Policy, 38*, 7457–7465.
4. Cheng, J. S. (2000). A survey of development of liquefied natural gas industry in the world. *Indian Natural Gas, 20*, 101–105.
5. Richman, R. H., & McNaughton, W. P. (1997). A metallurgical approach to improved cavitation-erosion resistance. *Journal of Materials Engineering and Performance, 6*, 633–641.
6. Santa, J. F., Blanco, J. A., Giraldo, J. E., & Toro, A. (2011). Cavitation erosion of martensitic and austenitic stainless steel welded coatings. *Wear, 271*, 1445–1453.
7. Okada, T., Hattori, S., & Shimizu, M. (1995). A fundamental study of cavitation erosion using a magnesium oxide single crystal (intensity and distribution of bubble collapse impact loads). *Wear, 186–187*, 437–443.
8. Kwok, C. T., Cheng, F. T., & Man, H. C. (2000). Synergistic effect of cavitation erosion and corrosion of various engineering alloys in 3.5 % NaCl solution. *Materials Science and Engineering A, 290*, 145–154.
9. Vyas, B., & Hansson, I. L. H. (1990). The cavitation erosion-corrosion of stainless steel. *Corrosion Science, 30*, 761–770.
10. Baek, J. H., Kim, Y. P., Kim, W. S., & Kho, Y. T. (2001). Fracture toughness and fatigue crack growth properties of the base metal and weld metal of a type 304 stainless steel pipeline for LNG transmission. *International Journal of Pressure Vessels Piping, 78*, 351–357.
11. Mukai, K., Hoshino, K., & Fujioka, T. (1979). Tensile and fatigue properties of austenitic stainless steels at LNG temperature. *Tetsu-To-Hagane(Journal of Iron Steel Institute of Japan), 65*, 1756–1765.

12. Zhang, L., Luo, K. Y., Lu, J. Z., Zhang, Y. K., Dai, F. Z., & Zhong, J. W. (2011). Effects of laser shock processing with different shocked paths on mechanical properties of laser welded ANSI 304 stainless steel joint. *Materials Science and Engineering A, 528*, 4652–4657.
13. Luo, K. Y., Lu, J. Z., Zhang, Y. K., Zhou, J. Z., Zhang, L. F., Dai, F. Z., et al. (2011). Effects of laser shock processing on mechanical properties and micro-structure of ANSI 304 austenitic stainless steel. *Materials Science and Engineering A, 528*, 4783–4788.
14. Zhang, Y. K., You, J., Lu, J. Z., Cui, C. Y., Jiang, Y. F., & Ren, X. D. (2010). Effects of laser shock processing on stress corrosion cracking susceptibility of AZ31B magnesium alloy. *Surface and Coatings Technology, 204*, 3947–3953.
15. Zhang, L., Lu, J. Z., Zhang, Y. K., Luo, K. Y., Zhong, J. W., Cui, C. Y., et al. (2011). Effects of different shocked paths on fatigue property of 7050-T7451 aluminum alloy during two-sided laser shock processing. *Materials and Design, 32*, 480–486.
16. Peyre, P., Carboni, C., Forget, P., Beranger, G., Lemaitre, C., & Stuart, D. (2007). Influence of thermal and mechanical surface modifications induced by laser shock processing on the initiation of corrosion pits in 316L stainless steel. *Journal of Materials Science, 42*, 6866–6877.
17. Sánchez-Santana, U., Rubio-González, C., Gomez-Rosas, G., Ocaña, J. L., Molpeceres, C., Porro, J., et al. (2006). Wear and friction of 6061-T6 aluminum alloy treated by laser shock processing. *Wear, 260*(7–8), 847–854.
18. Gomez-Rosas, G., Rubio-Gonzalez, C., Ocaña, J. L., Molpeceres, C., Porro, J. A., Morales, M., et al. (2010). Laser Shock Processing of 6061-T6 Al alloy with 1064 nm and 532 nm wavelengths. *Applied Surface Science, 256*(20), 5828–5831.
19. Zhang, H., & Yu, C. Y. (1998). Laser shock processing of 2024–T62 aluminum alloy. *Materials Science and Engineering A, 257*, 322–327.
20. Rubio-González, C., Gomez-Rosas, G., Ocaña, J. L., Molpeceres, C., Banderas, A., Porro, J., et al. (2006). Effect of an absorbent overlay on the residual stress field induced by laser shock processing on aluminum samples. *Applied Surface Science, 252*(18), 6201–6205.
21. Peyre, P., & Fabbro, R. (1995). Laser shock processing: A review of the physics and applications. *Optique Quantum Electron, 27*, 1213–1229.
22. Zhang, Y. K., Gu, Y. Y., Zhang, X. Q., & Zhou, J. (2006). Study of mechanism of overlay acting on laser shock waves. *Journal of Applied Physics, 100*, 1–103517.
23. ASTM G5-94. (1995). Standard test method of potentiodynamic test, American Society of Testing and Materials, Philadelphia, PA.
24. Al-Obaid, Y. F. (1995). The effect of shot peening on stress corrosion cracking behaviour of 2205-duplex stainless steel. *Engineering Fracture Mechanics, 51*, 19–25.
25. Lu, J. Z., Luo, K. Y., Zhang, Y. K., Cui, C. Y., Sun, G. F., Zhou, J. Z., et al. (2010). Grain refinement of LY2 aluminum alloy induced by ultra-high plastic strain during multiple laser shock processing impacts. *Acta Materialia, 58*, 3984–3994.
26. Harold, L., & Michael, R. H. (2008). The effects of laser peening on high-cycle fatigue in 7085-T7651 aluminum alloy. *Materials Science and Engineering A, 477*, 208–216.
27. Sano, Y., Obata, M., Kubo, T., Mukai, N., Yoda, M., Masaki, K., et al. (2006). Retardation of crack initiation and growth in austenitic stainless steels by laser peening without protective coating. *Materials Science and Engineering A, 417*, 334–340.
28. Lu, J. Z., Luo, K. Y., Zhang, Y. K., Sun, G. F., Gu, Y. Y., Zhou, J. Z., et al. (2010). Grain refinement mechanism of multiple laser shock processing impacts on ANSI 304 stainless steel. *Acta Materialia, 58*, 5354–5362.
29. Takakuwa, O., & Soyama, H. (2012). Using an indentation test to evaluate the effect of cavitation peening on the invasion of the surface of austenitic stainless steel by hydrogen. *Surface and Coatings Technology, 206*, 3747–3750.
30. Hsieh, C. C., Lin, D. Y., Chen, M. C., & Wu, W. (2008). Precipitation and strengthening behavior of massive $\delta$-ferrite in dissimilar stainless steels during massive phase transformation. *Materials Science and Engineering A, 477*, 328–333.

31. Xu, P. G., Tomota, Y., Lukáš, P., Muránsky, O., & Adachi, Y. (2006). Austenite-to-ferrite transformation in low alloy steels during thermomechanically controlled process studied by in situ neutron diffraction. *Materials Science and Engineering A, 435–436*, 46–53.
32. Fan, J. W., Xie, R. P., Wang, Y. C., & Liu, T. H. (2012). Fatigue behavior of plain C-Mn steel plates with fine grained ferrite in surface layers. *Materials Science and Engineering A, 539*, 154–162.
33. Saeidi, N., & Ekrami, A. (2010). Impact properties of tempered bainite-ferrite dual phase steels. *Materials Science and Engineering A, 527*, 5575–5581.
34. Arif, A. F. M. (2003). Numerical prediction of plastic deformation and residual stresses induced by laser shock processing. *Journal of Materials Processing Technology, 136*, 120–138.
35. Zhang, Y. K., Zhang, L., Luo, K. Y., Sun, G. F., Lu, J. Z., Dai, F. Z., et al. (2012). Effects of laser shock processing on mechanical properties of laser welded ANSI 304 stainless steel joint. *Chinese Journal of Mechanical Engineering, 25*, 285–292.
36. Zhang, Y. K., Lu, J. Z., Ren, X. D., Yao, H. B., & Yao, H. X. (2009). Effect of laser shock processing on the mechanical properties and fatigue lives of the turbojet engine blades manufactured by LY2 aluminum alloy. *Materials and Design, 30*(5), 1697–1703.
37. Brunatto, S. F., Allenstein, A. N., Allenstein, C. L., & Buschinelli, A. J. (2012). Cavitation erosion behaviour of niobium. *Wear, 274–275*, 220–228.
38. Ryl, J., Darowicki, K., & Slepski, P. (2011). Evaluation of cavitation erosion–corrosion degradation of mild steel by means of dynamic impedance spectroscopy in galvanostatic mode. *Corrosion Science, 53*, 1873–1879.
39. Li, X. Y., Yan, Y. G., Ma, L., Xu, Z. M., & Li, J. G. (2004). Cavitation erosion and corrosion behavior of copper-manganese-aluminum alloy weldment. *Materials Science and Engineering A, 382*, 82–89.
40. Boven, G. V., Chen, W., & Rogge, R. (2007). The role of residual stress in neutral pH stress corrosion cracking of pipeline steels. Part I: Pitting and cracking occurrence. *Acta Materialia, 55*, 29–42.
41. Turnbull, A., Mingard, K., Lord, J. D., Roebuck, B., Tice, D. R., Mottershead, K. J., et al. (2011). Sensitivity of stress corrosion cracking of stainless steel to surface machining and grinding procedure. *Corrosion Science, 53*, 3398–3415.
42. Blasco-Tamarit, E., García-García, D. M., & Antón, J. G. (2011). Imposed potential measurements to evaluate the pitting corrosion resistance and the galvanic behaviour of a highly alloyed austenitic stainless steel and its weldment in a LiBr solution at temperatures up to 150 °C. *Corrosion Science, 53*, 784–795.
43. Wang, Y. H., Wang, W., Liu, Y. Y., Zhong, L., & Wang, J. (2011). Study of localized corrosion of 304 stainless steel under chloride solution droplets using the wire beam electrode. *Corrosion Science, 53*, 2963–2968.
44. Flitt, H. J., & Schweinsberg, D. P. (2005). Evaluation of corrosion rate from polarisation curves not exhibiting a Tafel region. *Corrosion Science, 47*, 3034–3052.
45. Ramya, S., Anita, T., Shaikh, H., & Dayal, R. K. (2010). Laser Raman microscopic studies of passive films formed on type 316LN stainless steels during pitting in chloride solution. *Corrosion Science, 52*, 2114–2121.
46. Sulyma, C. M., & Roy, D. (2010). Voltammetric current oscillations due to general and pitting corrosion of tantalum: Implications for electrochemical-mechanical planarization. *Corrosion Science, 52*, 3086–3098.
47. Aquino, J. M., Della, C. A., & Kuri, S. E. (2009). Intergranular corrosion susceptibility in supermartensitic stainless steel weldments. *Corrosion Science, 51*, 2316–2323.
48. Kwok, C. T., Lo, K. H., Chan, W. K., Cheng, F. T., & Man, H. C. (2011). Effect of laser surface melting on intergranular corrosion behaviour of aged austenitic and duplex stainless steels. *Corrosion Science, 53*, 1581–1591.
49. Cui, Z. D., Man, H. C., Cheng, F. T., & Yue, T. M. (2003). Cavitation erosion-corrosion characteristics of laser surface modified NiTi shape memory alloy. *Surface and Coatings Technology, 162*, 147–153.

50. Tovar, R. S., Montanes, M. T., & Anton, J. G. (2010). Effect of different micro-plasma arc welding (MPAW) processes on the corrosion of AISI 316L SS tubes in LiBr and $H_3PO_4$ solutions under flowing conditions. *Corrosion Science, 52*, 1508–1519.
51. Chang, J. T., Yeh, C. H., He, J. L., & Chen, K. C. (2003). Cavitation erosion and corrosion behavior of Ni-Al intermetallic coatings. *Wear, 255*, 162–169.
52. Kwok, C. T., Cheng, F. T., & Man, H. C. (2006). Cavitation erosion and corrosion behaviors of laser-aluminized mild steel. *Surface and Coatings Technology, 200*, 3544–3552.
53. Kwok, C. T., Man, H. C., & Cheng, F. T. (2001). Cavitation erosion-corrosion behaviour of laser surface alloyed AISI 1050 mild steel using NiCrSiB. *Materials Science and Engineering A, 303*, 250–261.
54. García-García, D. M., García-Antón, J., Igual-Muñoz, A., & Blasco-Tamarit, E. (2006). Effect of cavitation on the corrosion behaviour of welded and non-welded duplex stainless steel in aqueous LiBr solutions. *Corrosion Science, 48*, 2380–2405.
55. Sánchez-Tovar, R., Montañés, M. T., & García-Antón, J. (2011). Effect of the micro-plasma arc welding technique on the microstructure and pitting corrosion of AISI 316L stainless steels in heavy LiBr brines. *Corrosion Science, 53*, 2598–2610.
56. Kwok, C. T., Man, H. C., & Cheng, F. T. (1998). Cavitation erosion and pitting corrosion of laser surface melted stainless steels. *Surface and Coatings Technology, 99*, 295–304.
57. Smialowska, Z. S. (1999). Pitting corrosion of aluminum. *Corrosion Science, 41*, 1743–1767.

# About the Authors

**Prof. Yongkang Zhang** is working at School of Mechanical Engineering, Southeast University, People's Republic of China. He received his Ph.D degree at Nanjing University of Aeronautics and Astronautics (P.R. China) in 1995. He subsequently completed his post-doctoral education at Nanjing University (P.R. China) in 1997. His research focuses on laser shock processing/peening, laser-ultrasonic detection, laser surface treatment and laser compound processing.
E-mail: zykseu@163.com

**Dr. Jinzhong Lu** is an associate professor at School of Mechanical Engineering, Jiangsu University, People's Republic of China. He received his Ph.D degree at Jiangsu University (P.R. China) in 2010. His research includes laser shock processing/peening, and non-traditional manufacturing technology based on laser technology.
E-mail: jxscience@ujs.edu.cn; blueesky2012@gmail.com

**Dr. Kaiyu Luo** is currently an associate professor at School of Mechanical Engineering, Jiangsu University, People's Republic of China. She received her Ph.D degree at Jiangsu University (P.R. China) in 2012. Her research includes laser shock processing/peening, laser cladding and laser compound processing.
E-mail: luoteacher@126.com

Y. Zhang et al., *Laser Shock Processing of FCC Metals*,
Springer Series in Materials Science 179, DOI: 10.1007/978-3-642-35674-2,

# Index



Y. Zhang et al., *Laser Shock Processing of FCC Metals*,
Springer Series in Materials Science 179, DOI: 10.1007/978-3-642-35674-2,

MIX
Papier aus verantwortungsvollen Quellen
Paper from responsible sources
FSC® C105338

If you have any concerns about our products,
you can contact us on
**ProductSafety@springernature.com**

In case Publisher is established outside the EU,
the EU authorized representative is:
**Springer Nature Customer Service Center GmbH**
**Europaplatz 3, 69115 Heidelberg, Germany**

Printed by Libri Plureos GmbH
in Hamburg, Germany